Emerging Trends in Mechatronics

Series Editor

Aydin Azizi, Oxford, UK

This book series helps understand how Mechatronics will impact the practice and research of developing advanced techniques to model, control and optimize complex systems. It presents recent state of advances in mechatronics and related technologies including: Automatic Control, Robotics, Agent Based systems, Smart Manufacturing and Industry 4.0. The selected topics give an overview of the state-of-the-art and present new research results and prospects of the future development in the interdisciplinary field of mechatronic systems. The book series will provide up-to-date and useful knowledge for researchers and engineers involved in mechatronics and related fields.

Within the above dimensions, the scope of the book series welcomes high quality original research and review articles that cover a broad range of topics related to theoretical and applied aspects of modern mechatronics. Potential topics include, but are not limited to the following:

Controller design

Many mechatronic systems have challenging control difficulties due to the high nonlinear structures and time-varying dynamic behaviors. In addition to these, there are external disturbances, which cannot be predicted and change according to uncontrolled working environments. Conventional controllers are insufficient to solve the problems and to compensate the environmental disturbances, therefore designing efficient controllers for complex systems has a great importance in the current book series.

Sensor Design & Data Collection Approaches

Essential part of a mechatronics system is the measurement system that senses the variations in the physical parameters, such as temperature, pressure, displacement, and so on, and converts it to voltage or current. In mechatronic systems, accurate measurement of the dynamic variables plays a vital role for the actuators to function properly, therefore on of the aims of this book series, to focus on understanding what should be done to design accurate sensors and efficient data collection methods to improve the performance of mechatronic systems.

Optimization Techniques

Designing a complex mechatronic system involves multiple variables, objectives, constraints, and evaluation criteria as well as their nonlinearly coupled relationships. One of the challenging research areas of this book series is utilizing optimization techniques to design an optimal mechatronic system.

Artificial Intelligence

From the point of view of dynamics, mechatronic systems can be characterized by model uncertainties, high nonlinearities, complicated coupling, and stringent performance requirements, among others. To deal with these facts, this book series would like to focus on investigating and utilizing Artificial Intelligence techniques such as Neural Networks, Fuzzy logic, Neuro-fuzzy, evolutionary techniques etc. to model, design and optimize complex mechatronic systems.

Robotics

In the field of Robotics, there is a need of focusing on kinematic, dynamic, control and optimization of innovative robotic systems. This book series can introduce new and novel robot designs such as hybrid (serial-parallel) redundant manipulators that have a high potential of application in special fields like space robots. State of the art researches on selection of optimum kinematic solutions and path planning based on such an optimization, dynamic analysis and new control methods developments and application of artificial intelligence in this field can be published in this book series.

Smart manufacturing

Application of networked sensors and machinery equipment using internet and intranet in order to monitor and automation of the manufacturing process and improve the performance efficiency and is a point of interest. Advanced robotics, multi input-multi output control approaches combined with the artificial intelligence and optimization techniques, big data processing approaches, multi-scale modelling and simulation of dynamic sub-systems etc. are the hot topics that will be focused in this book series.

AGVs

Automated Guided Vehicle (AGVs) are going to be very popular is industry logistics, surveillance, defense and social applications in order to make processes safer and more cost efficient. Innovative AGVs designs for various missions, Robot Operating System usage on AGV control, Indoor and outdoor navigation and control algorithms, intelligent path planning techniques, sensor fusion, simultaneous application of GPS, INS and offline maps are the topics that are planned to be PUBLISHED IN THIS BOOK SERIES.

Masoomeh Mirrashid
Danial Jahed Armaghani • Aydin Azizi
Editors

Intelligent Predictive Systems

AI and Machine Learning in Engineering

Springer

Editors
Masoomeh Mirrashid
College of Engineering
Abu Dhabi University
Al Ain, United Arab Emirates

Danial Jahed Armaghani
School of Civil and Environmental Engineering
University of Technology Sydney
Sydney, NSW, Australia

Aydin Azizi
School of Engineering, Computing and Mathematics
Oxford Brookes University
Oxford, UK

ISSN 2731-4855 ISSN 2731-4863 (electronic)
Emerging Trends in Mechatronics
ISBN 978-981-95-7765-1 ISBN 978-981-95-7766-8 (eBook)
https://doi.org/10.1007/978-981-95-7766-8

This Springer imprint is published by the registered company Springer Nature Singapore Pte Ltd.
The registered company address is: 152 Beach Road, #21-01/04 Gateway East, Singapore 189721, Singapore

Preface

New engineering methods are being developed and replaced with predictive systems based on Artificial Intelligence (AI) and Machine Learning (ML) technologies. This book, *Intelligent Predictive Systems: AI and Machine Learning in Engineering*, aims to capture and share some of these technologies in overcoming challenges. The chapters in this volume illustrate the diverse and detailed insights the field offers. This exploration is grounded in the fundamentals, with a devoted chapter on geotechnical engineering in which, the readers encounter a set of models moving beyond laboratory evaluations predicting the California Bearing Ratio (CBR). The predictive models are also supplemented with review chapters and synthesize the progress in ML predicting foundation and ground settlement with an overview and future research directions. One of the chapters illustrates the use of satellite photos coupled with an ML algorithm to gauge the trend of Lake Urmia's gradual drying and provides key details for the effective management of water resources. Another of the chapters describes the use of AI to improve the operational efficiency of public buses. The importance of AI in the efficient and sustainable use of materials is another theme in the book. The use of ML to predict the strength of concrete containing fly ash and marble slurry, and the enhancement of marine clay with recycled tiles, constitutes another chapter. Such construction prediction techniques and the use of industrial waste for concrete show the potential of ML for construction circular economy.

Al Ain, United Arab Emirates — Masoomeh Mirrashid
Sydney, NSW, Australia — Danial Jahed Armaghani
Oxford, UK — Aydin Azizi

Contents

Chapter 1
Machine Learning for Sustainable Concrete: Predictive Approaches Using Industrial Waste Materials

Muhammad Shahid Khan, Muhammad Khubaib, and Shaida Muhammad

Abstract Fly ash (FA) has emerged as a promising additional cementitious ingredient in the concrete industry, with considerable benefits in terms of sustainability and performance of concrete. At the same time, the rapid increase in Marble Cutting Slurry Waste (MCSW) produced by ornamental stone firms has prompted major environmental issues about soil, air, and water contamination. The advantages of adding MCSW to concrete mixtures have been investigated in certain experiments by several researchers. Despite the known benefits of incorporating FA and MCSW in concrete, a data-driven understanding of its impact on compressive strength, particularly through Machine Learning (ML) models, remains underexplored. In this research work, an ML-based approach is used to forecast the compressive strength (CS) of concrete incorporating FA and MCSW, with the goal of improving resource use while addressing environmental concerns. A variety of machine learning techniques were applied, such as Decision Tree (DT), Random Forest (RF), Categorical Gradient Boosting (CatBoost), Extreme Gradient Boosting (XGBoost), and Light Gradient Boosting Machine (LightGBM). The dataset comprises 270 points. These points were compiled from available literature sources. The performance of the developed models is assessed through different statistical indicators. All five ML models did well, but XGBoost was the best in predicting CS. Notably, XGBoost showed the highest coefficient of determination (R^2) value of 0.97, signifying its greater prediction accuracy in contrast to DT, RF, LightGBM, and CatBoost. Moreover, the SHapley Additive exPlanations (SHAP) method is employed to inter-

M. S. Khan
School of Civil Engineering, Southwest Jiaotong University, Chengdu, China
e-mail: shahidkhan@my.swjtu.edu.cn

M. Khubaib (✉)
Department of Civil Engineering, University of Engineering and Technology Peshawar, Peshawar, Pakistan
e-mail: 22pwciv5809@uetpeshawar.edu.pk

S. Muhammad
Department of Computer Science, National University of Science and Technology Islamabad, Islamabad, Pakistan

M. Mirrashid et al. (eds.), *Intelligent Predictive Systems*, Emerging Trends in Mechatronics, https://doi.org/10.1007/978-981-95-7766-8_1

pret the predictions of the models. The findings provide a robust data-driven framework for optimizing eco-friendly concrete mix designs, contributing to sustainable construction practices and informed decision-making in material selection.

Keywords Concrete · Fly ash · Machine learning · Concrete mix design

Article Highlights

- Five machine learning models (DT, RF, XGBoost, LightGBM, and CatBoost) were developed and compared using a dataset to predict the compressive strength of concrete containing fly ash (FA) and marble cutting slurry waste (MCSW).
- A dataset of 270 experimental samples obtained from the literature was compiled and analyzed to ensure that the model training and validation were not affected by low sample statistics.
- The XGBoost model achieved the best predictive performance, yielding an R^2 value of 0.97 against other ML algorithms.

1.1 Introduction

The construction industry is currently the largest producer of greenhouse gases (GHG), accounting for approximately half of the overall emissions [1]. Rising carbon emissions are melting Arctic and Antarctic glaciers, posing serious global environmental risks [2]. Construction industry is currently the largest producer of greenhouse gases (GHG), accounting for about half of overall emissions [1]. Ordinary Portland Cement (OPC) is the main ingredient of concrete and a significant contributor to GHG emissions [3]. The production of Ordinary Portland Cement (OPC) contributes nearly 7% of global CO_2 emissions, with the calcination of calcium oxide (CaO) responsible for approximately half of these emissions [4]. According to the literature, Portland cement production is expected to increase from 4 to 6 billion metric tons per year by 2060 [5]. These findings highlight the urgent need for practical strategies to meet the growing demand for concrete while reducing energy consumption and CO_2 emissions [6, 7]. Using waste and recyclable resources in concrete is a practical and effective solution. This technique would not only meet the expanding need for concrete but also contribute to reducing the growing environmental risks.

FA is gaining popularity as a substitute for cement in concrete because of its sustainability advantages, reduced CO_2 emissions, and lower production costs compared to OPC [8]. The utilization of calcined clay, slag, FA, or volcanic ash in concrete presents numerous potential benefits, such as enhanced sustainability and decreased expenses [9–11]. FA serves as a substitute for cement and improves the bonding properties and mineralogical mixture of wet and aged concrete for construction [12]. The utilization of FA significantly reduces the volume of water required for concrete production. It can enhance the mechanical properties of concrete, affecting both its compressive and split-tensile strengths [13].

The disposal of MCSW has become a serious concern for both the environment and human health [14]. Approximately 25% of the original mass is lost when marble is cut, leaving behind marble slurry that pollutes groundwater and endangers biodiversity [14]. MCSW not only improves the physical properties of concrete, but it also offers a sustainable way to dispose of waste. For example, the use of 20% MCSW as a substitute for sand in concrete demonstrated superior mechanical and durability characteristics relative to the standard mix [15].

Kavya et al. substituted cement with 20% FA and sand with MCSW at varying doses (0%, 10%, 20%, and 30%) [16]. They found that replacing 20% of the sand with MCSW enhanced both the workability and mechanical qualities of the concrete [16]. According to a study using 10% FA and 40% MCSW, the CS of concrete increased by 16.21% compared to the standard mix [17]. The use of MCSW and FA in concrete lowers the expense of paver bricks, making it both inexpensive and environmentally friendly [18]. Similarly, another study showed that utilizing MCSW and FA increases the mechanical characteristics of concrete [19]. Utilizing MCSW and fly ash at various proportions in concrete offers affordable and environmentally friendly high-strength concrete while minimizing the need for large amounts of cement and lowering CO_2 emissions.

The main goal of incorporating waste material into concrete production is to reduce CO_2 emissions during cement manufacturing, thereby decreasing the demand for cement, reducing its environmental impact, and conserving natural resources. Therefore, increasing the marble dust and fly ash replacement ratios can help achieve environmental objectives. Previous experimental studies on using marble waste and fly ash in concrete production indicate that their use as a replacement for cement and fine aggregate at various proportions should be limited to a specific proportion, as the compressive strength changes with different proportions [20–23]. Despite considerable research aimed at enabling wide-scale partial utilization of fine aggregates and binders in concrete incorporating MCSW and FA, its implementation in mainstream usage has been hampered by a lack of quantitative research and ambiguities in laboratory findings, mostly due to the narrow range of studied concrete mixes and factors. A clear and comprehensive understanding of the optimal method and dosage for incorporating MCSW and FA to maximize the engineering properties of concrete remains lacking. Owing to insufficient experimental data and the absence of a comprehensive model that accurately reflects the behavior; the reliability of earlier findings often shows contradictory relationships between concrete mix components and mechanical strength [24]. Notably, fluctuations in experimental settings make it impossible to standardize and produce common concrete mixing formulations, including MCSW and FA.

Therefore, ML and Data Mining (DM) approaches can provide cost and time-efficient methods to model this behavior. Consequently, data-based computational modeling could provide an additional arena for enhancing the usage and applications of MCSW and FA in the concrete industry, supplementing the findings of previous experimental studies.

Numerous studies have explored the use of ML in concrete mix design and compressive strength prediction [25–32], demonstrating the ability of ML methods in this domain. While advanced ML models have gained significant attention from researchers, their application in predicting the behavior of concrete containing MCSW and FA remains to be explored. Many ML algorithms, particularly deep neural networks, require significant computational power and are time-intensive. Furthermore, developing a more complex model can lead to overfitting, which reduces its generalizability, whereas the main goal is to develop straightforward solutions for practical future on-site applications. ML models require large amounts of data; therefore, supplying extensive and well-distributed datasets across all dimensions can improve their generalizability. Despite the well-known benefits of integrating MCSW and FA into concrete, data-driven knowledge of their impact on CS, particularly through ML models, is currently unexplored. The addition of ingredients such as MCSW and FA complicates the prediction of the CS of concrete owing to their diverse impacts on the physical and chemical attributes of concrete. Machine learning methods, with their ability to handle nonlinearities and interactions within datasets, provide a potential solution for untangling this complexity.

This study presents ML model development and conducts a thorough evaluation of ML models for CS prediction of concrete incorporating MCSW and FA in various proportions. No prior study has been systematically compared boosting-based machine learning algorithms (XGBoost, LightGBM, CatBoost) for forecasting the compressive strength of concrete incorporating MCSW and FA.

1.2 Overview of the Applied Machine Learning Methods

This section discusses the theoretical foundations and procedures of the machine learning techniques adopted in this study.

1.2.1 Decision Tree (DT)

Decision tree classifiers are commonly regarded as one of the most common methods for presenting classifiers in classification of data. Scholars from various disciplines and backgrounds have investigated the difficulty of expanding decision trees based on available data, including machine learning, pattern identification, and statistics [33]. Decision tree learning stands out as a highly successful algorithm because of its appealing characteristics, including simplicity, ease of understanding, limited parameters, and the capability to manage data of mixed types. This approach involves inducing a DT from a collection of labeled training examples, each depicted by a combination of feature values and corresponding target labels [34]. The primary advantage of decision trees is their capability to develop rule-based classifiers that are easily understandable to humans. However, they also have limitations, such

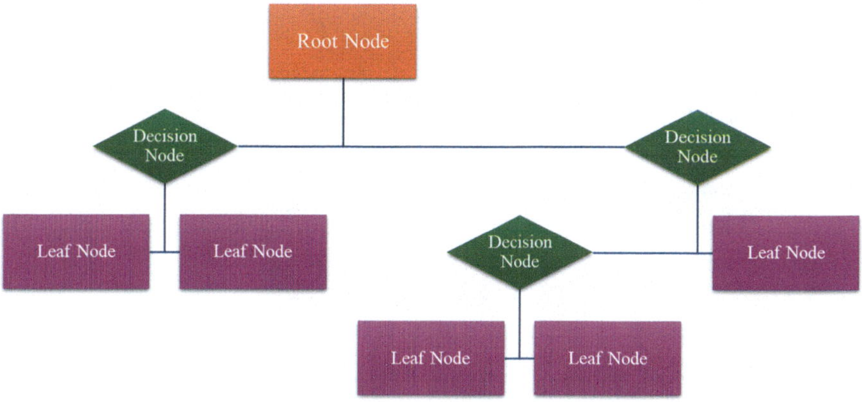

Fig. 1.1 Illustration of DT

as the requirement to organize all numerical attributes to determine node splitting locations [35]. A decision tree represents a hierarchical structure in which decision nodes represent tests applied to the configurations of input data and leaf nodes signify groups assigned to these configurations. As the input patterns traverse the tree, they are guided by these tests to produce the appropriate output. Decision tree algorithms find application across diverse fields owing to their versatility and effectiveness [36], and the DT is classified in Fig. 1.1.

1.2.2 Random Forest (RF)

Random forests comprise numerous tree predictors, where every tree depends on scores from an arbitrary vector sampled separately and uniformly throughout all trees in the forest. As the number of trees within the forest increases, the prediction errors of the limitations of Random Forest methods increase [37]. Typically, each machine learner's vote carries equal weight in the Random Forest algorithm. However, in later revisions by Breiman, this approach was adapted to handle both weighted and unweighted voting schemes. Ultimately, the forest selects the prediction based on the highest vote count among individual classifiers [38]. Random Forest has gained widespread adoption in classification and prediction tasks and has also found applications in regression. Compared to traditional algorithms, Random Forest exhibits numerous favorable qualities [39]. Random Forest has emerged as a highly effective and resilient algorithm capable of addressing feature selection challenges, even with many variables. It efficiently handles tasks such as the imputation of missing data, classification, and regression. Additionally, Random Forest demonstrates proficiency in managing outliers and noisy data [40], and a DF illustration is shown in Fig. 1.2.

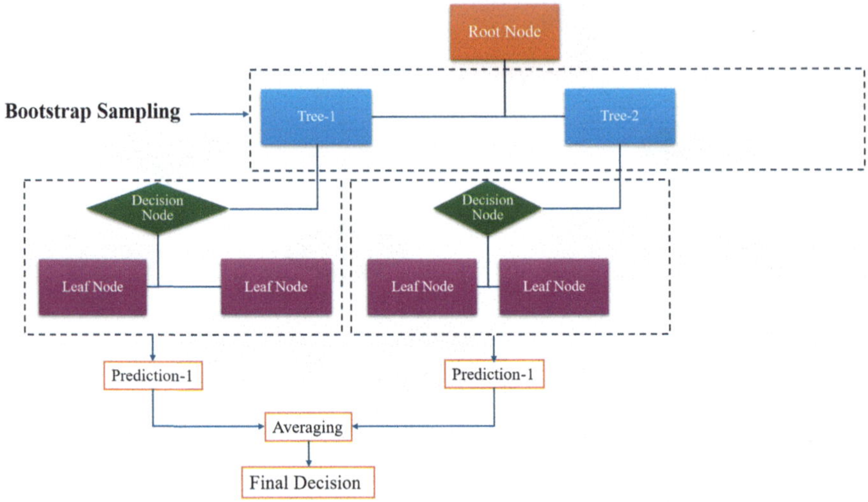

Fig. 1.2 Illustration of RF

1.2.3 XGBoost Algorithm (Extreme Gradient Boosting)

XGBoost is an ensemble ML that is effectively applied to classification and regression model prediction. XGBoost has demonstrated outstanding efficiency in tackling complicated nonlinear problems, owing to its unique method for preventing overfitting through careful hyperparameter optimization and good missing value handling [41, 42]. XGBoost uses gradient boosting to maximize its prediction power [42, 43]. Furthermore, this approach aims to reduce the difference between the real and anticipated values. It is accomplished through repeatedly generating new decision trees meant to capture the flaws (residuals) of earlier predictions, as illustrated in Fig. 1.3. The XGBoost iterative method improves accuracy and prediction skills, thereby highlighting its value for machine learning and data analysis [44, 45].

1.2.4 LightGBM Model: Light Gradient Boosting Machine

LightGBM is a gradient boosting architecture that employs tree-based learning techniques. LightGBM outperformed previous approaches in terms of accuracy and memory efficiency, making it ideal for large datasets. This approach uses techniques such as unique gradient-based one-side sampling (GOSS) and feature bundling to handle large datasets and attributes without overfitting [46]. A unique characteristic of LightGBM is the application of the histogram method combined with a leaf-wise tree growth procedure. The program separates discretizing floating-point values into

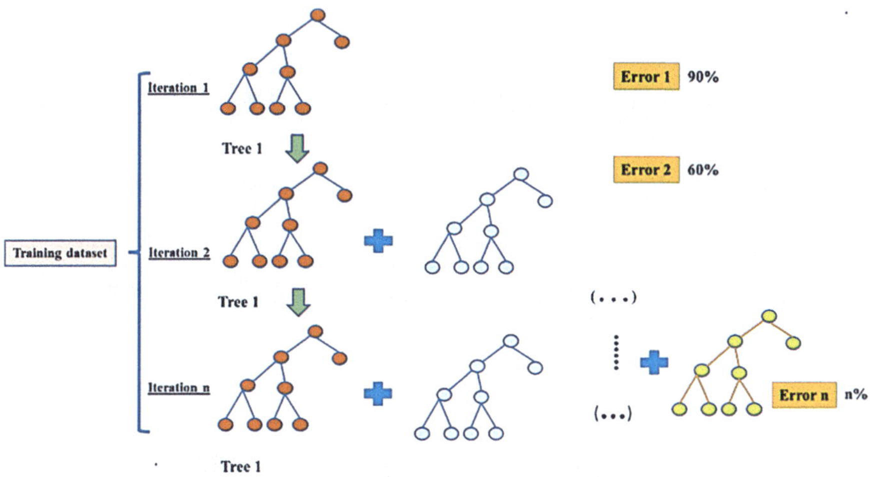

Fig. 1.3 Schematic representation of XGBoost approach [49]

smaller intervals and creates a histogram based on their width. It retrieves important statistics from the histogram, including the sum of gradients and quantity of samples, during the initial data analysis. This approach uses discrete values to determine appropriate segmentation points, which reduces storage and computational expenses [47]. Figure 1.4 shows an illustration of the LightGBM method. Metakaolin-based high-strength concrete exhibits significantly improved resistance to chloride penetration when evaluated using advanced machine learning models. LightGBM and XGBoost, supported by SHAP analysis, demonstrated strong predictive capability for rapid chloride penetration resistance [48].

1.2.5 CatBoost Algorithm (Categorical Gradient Boosting)

CatBoost is a powerful ML algorithm that enhances the classic gradient boosting algorithm to efficiently deal with variables of different categories. Conventional gradient-boosted decision trees depend on numerical attributes; however, categorical inputs require preprocessing. However, categorical gradient boosting, which is commonly used in frameworks such as CatBoost and LightGBM, makes this process more efficient [50]. Unlike traditional methods, CatBoost directly addresses the categorical attributes during tree formation. It employs an exclusive approach for categorical data, eliminating the need for long preprocessing procedures. This makes the workflow easier for practitioners and increases the model's prediction accuracy. CatBoost demonstrates the capacity to remove biases by implementing a dynamic boosting method, which improves the general performance of the CatBoost algorithm [51].

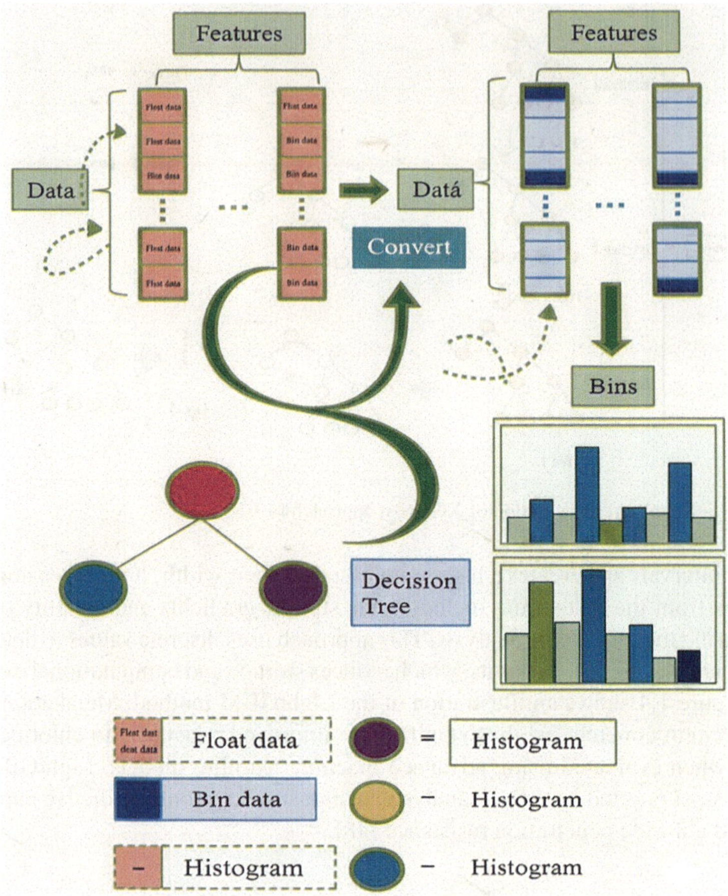

Fig. 1.4 Illustration of the LightGBM method

The method comprises methodologies for determining acceptable categorical variable splits while considering the unique properties of data that are non-ordinal. It overcomes issues such as attribute combinations and the absence of values that are uniquely designed for categorical data, enabling its effective application to real-world datasets with discrete features. Furthermore, CatBoost reduces the danger of overfitting using approaches such as regularization and tree pruning. This guarantees that the model correctly generalizes to new, previously unexplored datasets. The CatBoost procedure is illustrated in Fig. 1.5.

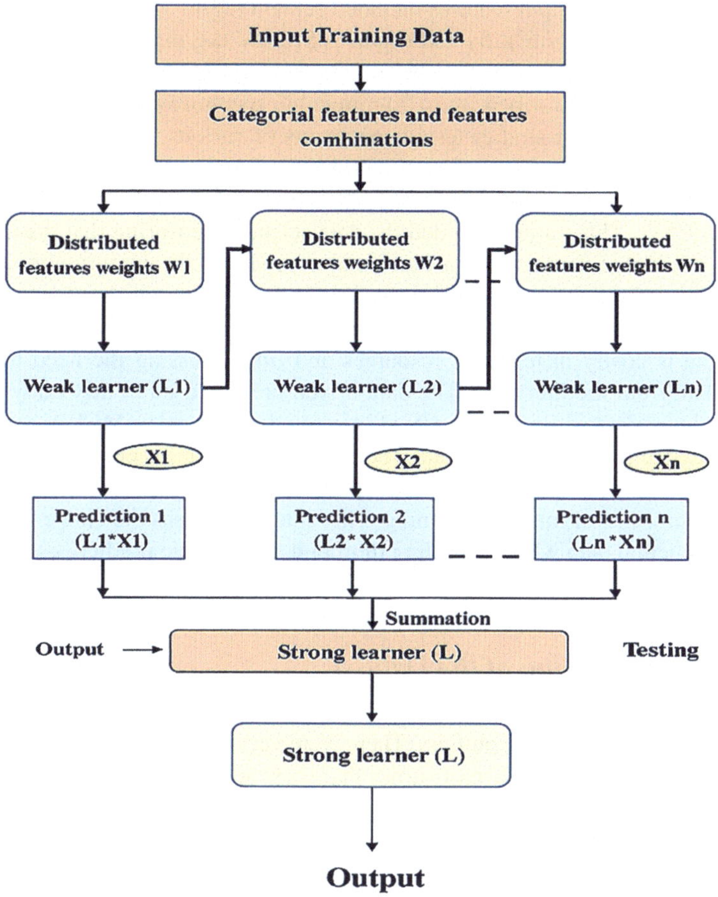

Fig. 1.5 Illustration of LightGBM method [49]

1.3 Research Methodology

1.3.1 Dataset Compilation

A thoroughly managed dataset is essential for ensuring the resilience and dependability of ML algorithms. The dataset for the present research was meticulously assembled to guarantee precision, relevance, and suitability to the research objectives. At the outset, an extensive examination of the existing literature was conducted using reliable sources, such as the Web of Science and Scopus. This guaranteed that the data collection was limited to high-quality peer-reviewed articles. The dataset contained 270 points from experiments conducted in prior research [20–23, 52–60]. Existing studies were reviewed to obtain precise information, which was gathered manually to maintain uniformity and ensure that critical details were not missed. The input features included cement (kg/m^3), coarse aggregate (kg/m^3), fine aggregate (kg/m^3), fly ash (kg/m^3), MCSW (kg/m^3), time in days, W/C, while the output or target was CS (MPa). To examine the CS of concrete, all gathered results were transformed into 100-mm cubic specimens using Neville's technique [61], as different studies tested specimens of various shapes and sizes. The primary objective of this research was to provide reliable and exact models for forecasting the compressive strength of concrete when cement and sand is replaced by FA and MCSW. This purpose guided the data selection, ensuring that the information was relevant to the factors that had a major impact on the compressive strength. These feature criteria were selected to enable precise predictions. Furthermore, conducting experiments to achieve reliable predictive results to obtain recently collected data is costly in terms of resources and time stressing the need to utilize current data from the literature. The data-driven method used in this study aims to improve the understanding of the CS of concrete incorporating MCSW and FA by utilizing existing experimental data. The studies' reliance on previous research is both practical and compatible with its purpose of examining the predictive and analytical possibilities of machine learning. This study addresses the challenge of limited experimental data while providing thorough and reliable predictive models.

1.3.2 Preprocessing of the Dataset

The dataset was checked for outliers. There were very few outliers in the features of fly ash and MCSW. The box plot shows that there were no outliers in the dataset. The dataset was also checked for missing values, and no missing values were found. Figure 1.6 shows the box plot used to check for outliers.

1.3.3 ML Algorithms Development

The prediction accuracy score of the ML algorithm is heavily reliant on proper hyperparameter tuning, as incorrect values might result in inaccurate targets by failing to lower the objective loss function values. In addition, training methods determine the performance and resilience of the ML model. During the training process, validation was used to achieve good accuracy and reduce model overfitting problems. Overfitting occurs when a model performs well while accurately capturing the training data but performs poorly on new inputs, with the testing dataset used to evaluate its generalization [62, 63]. Previous studies employed a five-fold cross-validation strategy, in which 20% of the dataset was randomly selected for testing in each fold to reliably evaluate model performance [64–67]. Numerous studies, such as DT [68, 69], RF [70, 71], LightGBM [72], XGBoost [68, 69], and CatBoost [73, 74] have utilized hyperparameter tuning to identify the optimal hyperparameter values for the models. In the present study, 10% data was used for validation, 20% for

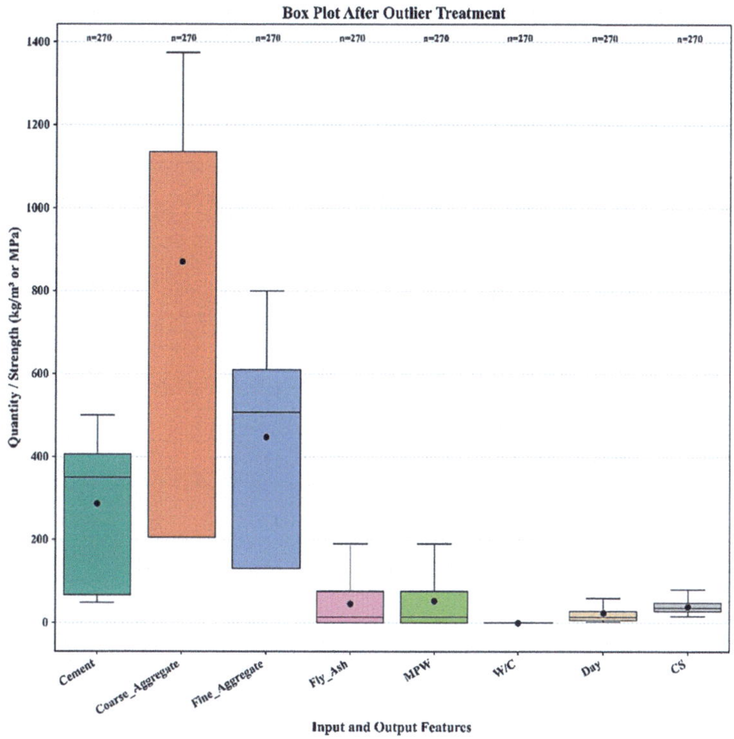

Fig. 1.6 Box plot to check for outliers

testing, and 70% for building and calibrating the machine learning models. The model's optimized hyperparameters, based on literature [68–78], are in Table 1.1, and the ML modeling framework is in Fig. 1.7.

1.3.4 Statistical Analysis

Every algorithm is subjected to a rigorous evaluation process in which multiple performance indicators are calculated to assess its accuracy and effectiveness [79]. A measure of how well the model explains the observed variance (R^2) reflects the capacity of the algorithm to explain the observed data, with values near 1 indicating high predictive ability. In concrete strength prediction experiments, the R^2 values often range from 0.80 to 0.98 [24]. The mean absolute deviation (MAE) represents the mean size of the errors, irrespective of their direction, with lower values signifying enhanced precision [80]. The RMSE, a metric indicating the average magnitude of errors in predictions, is often utilized for predicting the compressive strength of concrete because it is more sensitive to large errors owing to the squaring procedure [81]. The RSR, the ratio of the root mean square error to the standard deviation of

Table 1.1 Hyperparameter optimization for the employed algorithm

Algorithm	Hyperparameter	Search range	Best value
Decision Tree (DT)	Upper limit (max) depth	$1 \rightarrow \infty$	4
	Fewest required samples per split	≥2	2
	Fewest required samples per leaf	≥1	1
Random Forest (RF)	Quantity of trees (n_estimators)	≥10	200
	Features considered per split	√(features) → all features	√(features)
	Fewest required samples per split	≥2	2
	Maximum depth	≥1	36
	Fewest required samples per leaf	≥1	1
	Bootstrap sampling	–	Enabled (true)
XGBoost	Learning rate	(0, 1]	0.05
	Minimum child weight	≥0	1
	Subsample ratio	(0, 1]	0.8
	Maximum depth	≥1	3
	Boosting rounds	≥10	70
	Objective function	–	Reg: squared error
	Booster type	–	gbtree
	Column subsampling per tree	–	0.9
	L1 regularization (α)	–	0
	L2 regularization (λ)	–	0.1
LightGBM	Rate of model adjustment (learning rate)	(0, 1]	0.08
	Feature fraction	(0, 1]	0.6
	Maximum depth	≥1	12
	Number of iterations	≥10	200
	Minimum data in leaf	≥1	10
	Column subsampling per tree	–	0.9
	L1 regularization (α)	–	0
	L2 regularization (λ)	–	0.1
CatBoost	Maximum depth	1 → 5	5
	Learning rate	0.01 → 0.1	0.1
	L2 leaf regularization	1 → 20	1
	Number of trees (n_estimators)	150 → 300	300
	Column subsampling per level	0.5 → 1	1
	Additional max depth	≥1	8
	Minimum samples per split	≥2	2
	Minimum samples per leaf	≥1	5

the observed values normalizes the RMSE, making it easier to understand. RSR values ranging from 0.0 to 0.5 are usually deemed acceptable, and measurements close to 0 show great performance [82]. Engineering indices, such as the a10- and a20-index, which measure the proportion of predicted compressive strength values

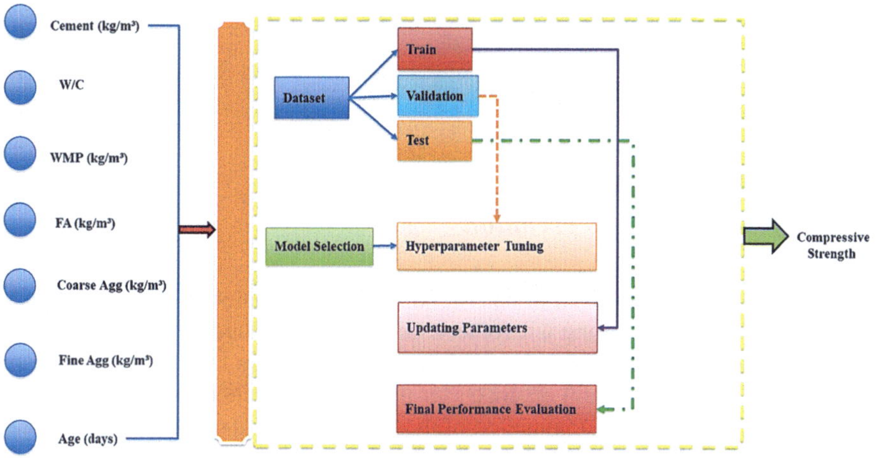

Fig. 1.7 Framework for ML modeling

that fall within ±10% and ±20% of the observed experimental values, respectively, are useful for evaluating model applicability in real-world engineering contexts. The reported values for the a10-index range from 60% to 80%, whereas the a20-index frequently reaches 85%, showing a significant alignment between the experimental and projected findings [83, 84]. Taken together, these criteria provide a thorough and comprehensive evaluation of the effectiveness of the ML model in forecasting the compressive strength of concrete. The formulas used to calculate these metrics are detailed in Eqs. (1.1)–(1.6).

$$R^2 = \frac{\left[\sum_{i=1}^{i=n}\left(Y_i - \overline{Y}_i\right)\left(¥_i - \overline{¥}_i\right)\right]^2}{\sum_{i=1}^{i=n}\left(Y_i - \overline{Y}_i\right)^2 \sum_{i=1}^{i=n}\left(¥_i - \overline{¥}_i\right)^2} \tag{1.1}$$

$$\mathrm{RMSE} = \sqrt{\frac{1}{N}\sum_{i=1}^{i=n}\left(Y_i - ¥_i\right)^2} \tag{1.2}$$

$$\mathrm{MAE} = \frac{1}{N}\sum_{i=1}^{i=n}\left|Y_i - ¥_i\right| \tag{1.3}$$

$$\mathrm{RSR} = \frac{\mathrm{RMSE}}{\mathrm{SD}} \tag{1.4}$$

$$\mathrm{a10\_index} = \frac{M10}{N} \tag{1.5}$$

$$\text{a20\_index} = \frac{M20}{N} \tag{1.6}$$

1.3.5 SHapley Additive exPlanations (SHAP) for Model Interpretation

In this study, SHAP was used to determine model explainability. Previous studies have shown that SHAP, along with other explainability methods, is frequently used to explain model predictions [85]. SHAP was proposed by Lundberg and Lee in 2017 as a model-independent approach for interpreting predictions [86]. The main idea behind SHAP is the computation of the SHapley contribution of each feature, which measures the incremental contribution of individual variables based on a conditional-expectation framework. SHAP utilizes SHapley values to provide a clear insight into the influence of each feature on model predictions, improving interpretability and transparency across various machine learning frameworks. Its flexibility makes SHAP an essential tool in explainable AI, allowing researchers and practitioners to thoroughly examine and comprehend the functioning of sophisticated models.

1.4 Results and Discussion

1.4.1 Examination of the Dataset

The dataset consists of 270 experimental records collected from existing literature. The input features consist of cement (kg/m^3), coarse aggregate (kg/m^3), fine aggregate (kg/m^3), fly ash (kg/m^3), MPW (kg/m^3), and curing period (days), while the target variable is the compressive strength (MPa). Table 1.2 presents a statistical overview of the dataset, incorporating the mean, standard deviation, maximum, and minimum values. Notably, the data were used in their original form without normalization during model development. The standard deviation varied notably across mixed ingredients, ranging from 431.49 for coarse aggregate to 0.07 for the W/C ratio. The W/C ratio exhibited a narrow range of 0.21, in contrast to the much wider range of 1169 observed for coarse aggregate. A correlation matrix, such as the one illustrated in Fig. 1.8, serves as an effective means of exploring correlations between variables within a dataset. The table displays the Pearson correlation values for each possible combination of variables in the dataset. These values capture both the intensity and the orientation of linear associations among the variables, spanning a range from −1 to 1. A value of −1 indicates a perfect inverse relationship, showing that the variables vary in opposite directions. A correlation score of 0 shows no

Table 1.2 Statistical summary of the dataset

	Cement (kg/m^3)	Coarse aggregate (kg/m^3)	Fine aggregate (kg/m^3)	Fly ash (kg/m^3)	MPW (kg/m^3)	W/C	Days	Compressive strength (MPa)
N	270	270	270	270	270	270	270	270
Mean	287.08	869.91	447.11	45.51	52.66	0.44	23.58	39.27
Std. dev.	155.31	431.49	226.43	62.68	69.6	0.07	17.77	14.4
Min	49	205	130	0	0	0.34	3	15.77
*Q*1	66.5	205	130	0	0	0.4	7	27.89
Median	350	1135	506.74	14.31	14.31	0.45	14	36.3
*Q*3	406	1135	610	76	76	0.55	28	48.95
Max	500	1374	800	190	190	0.55	59.5	80.54
Variance	24,119.92	186,181	51,271.78	3929.37	4844.75	0.01	315.65	207.32
Skewness	−0.65	−0.79	−0.3	1.37	1.1	0.33	0.83	0.56
Kurtosis	−1.29	−1.14	−1.28	0.46	−0.44	−1.22	−0.55	−0.48
Range	451	1169	670	190	190	0.21	56.5	64.77
Sum	77,510.86	234,875	120,718.9	12,287.49	14,218.25	120.09	6366.5	10,602.56

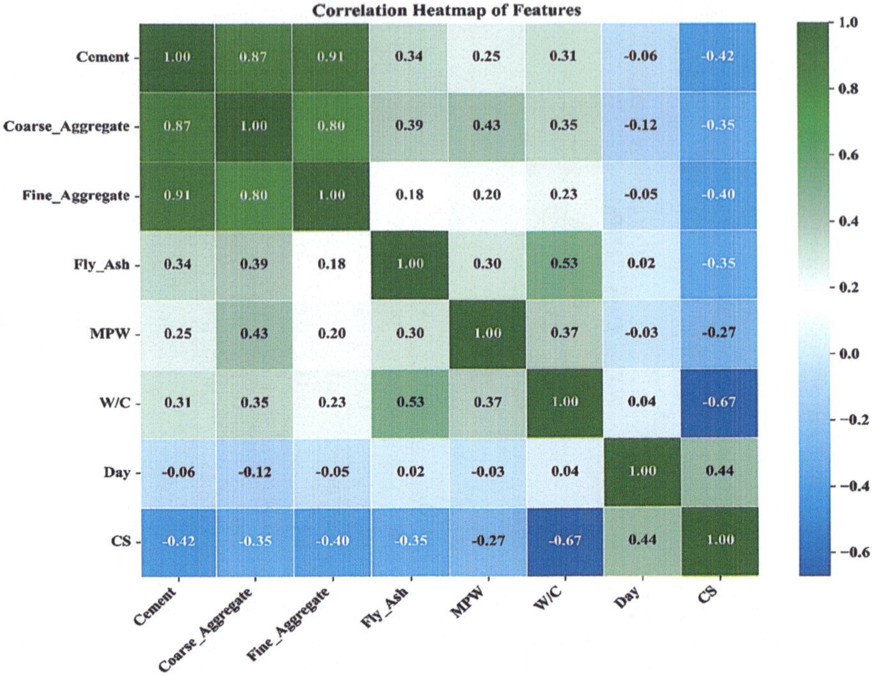

Fig. 1.8 Correlation matrix

association, whereas a score of 1 indicates a significant positive correlation, implying that the variables change in tandem. The matrix shows correlations between variables (−0.03 to −0.67); note that Pearson coefficients capture only linear relationships.

1.4.2 Regression Analysis of Employed Models

A set of five machine learning models DT, RF, CatBoost, LightGBM, and XGBoost was employed to estimate the compressive strength of concrete incorporating fly ash and MPW. The regression slope serves as a key metric for assessing the alignment between predicted outputs and true values in machine learning models. It was chosen due to its capability to offer both graphical and quantitative evaluation of prediction performance [87, 88]. If the predicted values match the observed values exactly, they fall along a straight line with a slope of one, indicating perfect agreement [87, 88]. Deviations from the ideal slope, on the other hand, highlight biases or consistent errors in the model's forecasts [87, 88]. A gradient (slope) near 1 shows predicted values from the model closely match the actual values, indicating proportionate consistency across the dataset [87, 88]. During training, the DT, RF, XGBoost, LightGBM, and CatBoost models achieved regression slopes of 0.82,

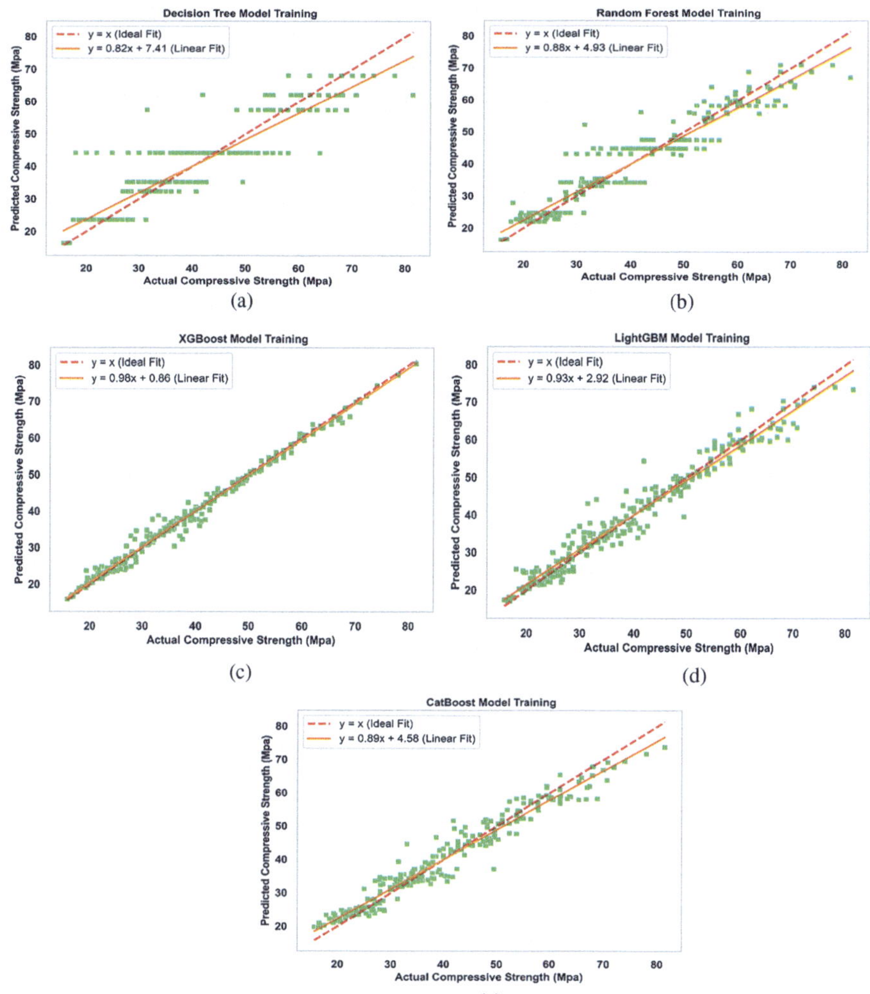

Fig. 1.9 Regression slopes of different models (**a**. Decision Tree, **b**. Random Forest, **c**. XGBoost, **d**. LightGBM, **e**. CatBoost) on the training data

0.88, 0.99, 0.93, and 0.94, respectively, as shown in Fig. 1.9. For the test dataset, the DT, RF, XGBoost, LightGBM, and CatBoost models produced regression line coefficients of 0.83, 0.86, 0.94, 0.89, and 0.89, respectively, as illustrated in Fig. 1.10. Similarly, for the validation dataset, the corresponding slopes were 0.80, 0.82, 0.88, 0.90, and 0.90, as shown in Fig. 1.11.

Compared to the other models, XGBoost achieved the greatest regression line coefficient observed for both training and test datasets, demonstrating superior precision in predicting the load-bearing capacity of concrete incorporating fly ash as a partial substitute for cement and marble powder waste as a replacement for sand.

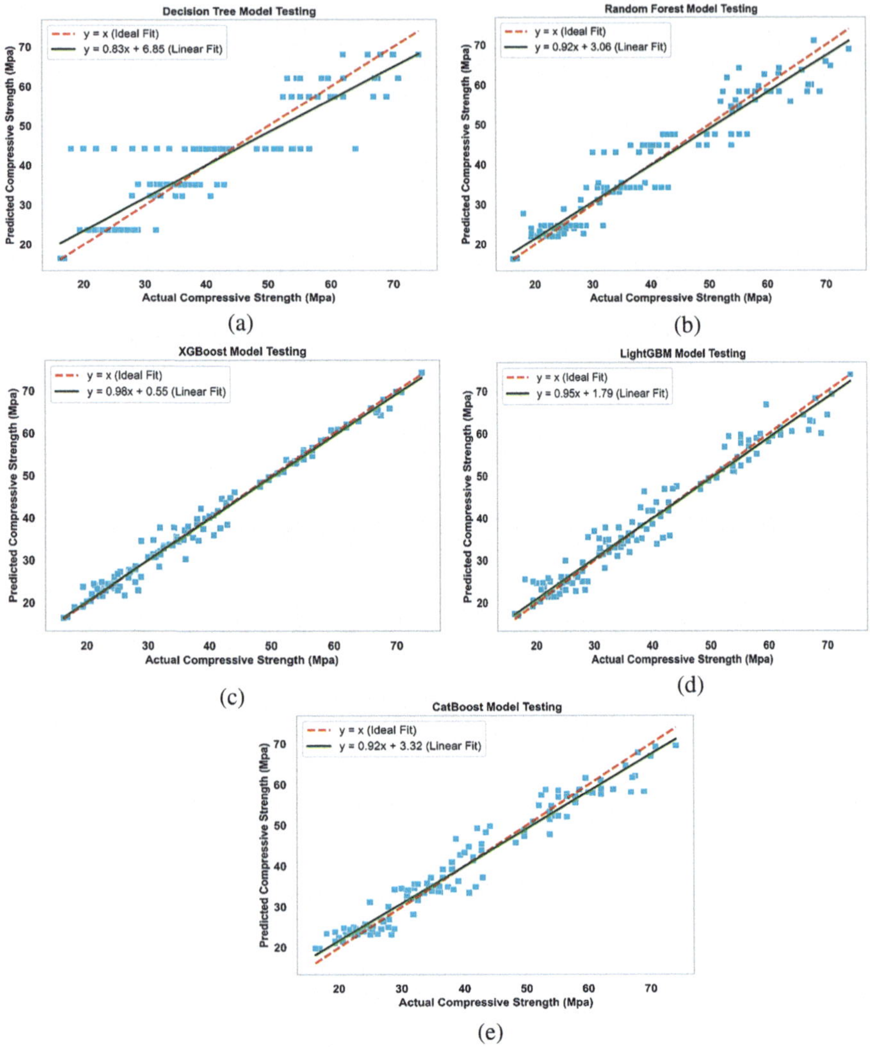

Fig. 1.10 Regression slope analysis of models (**a**. Decision Tree, **b**. Random Forest, **c**. XGBoost, **d**. LightGBM, **e**. CatBoost) on test data

1.4.3 Statistical Examination

The proposed models were rigorously evaluated using a range of statistical metrics, the outcomes of which are reported in Table 1.3. Overall, the deviation indices (MAE, RMSE) exhibit values approaching zero, while the reliability measures (*R*, a10-index) approach unity, demonstrating the superior accuracy of the ML models. The XGBoost model showed superior performance compared to DT, RF, LightGBM, and CatBoost, with *R* values of 1.0 for the training dataset and 0.97 for the testing

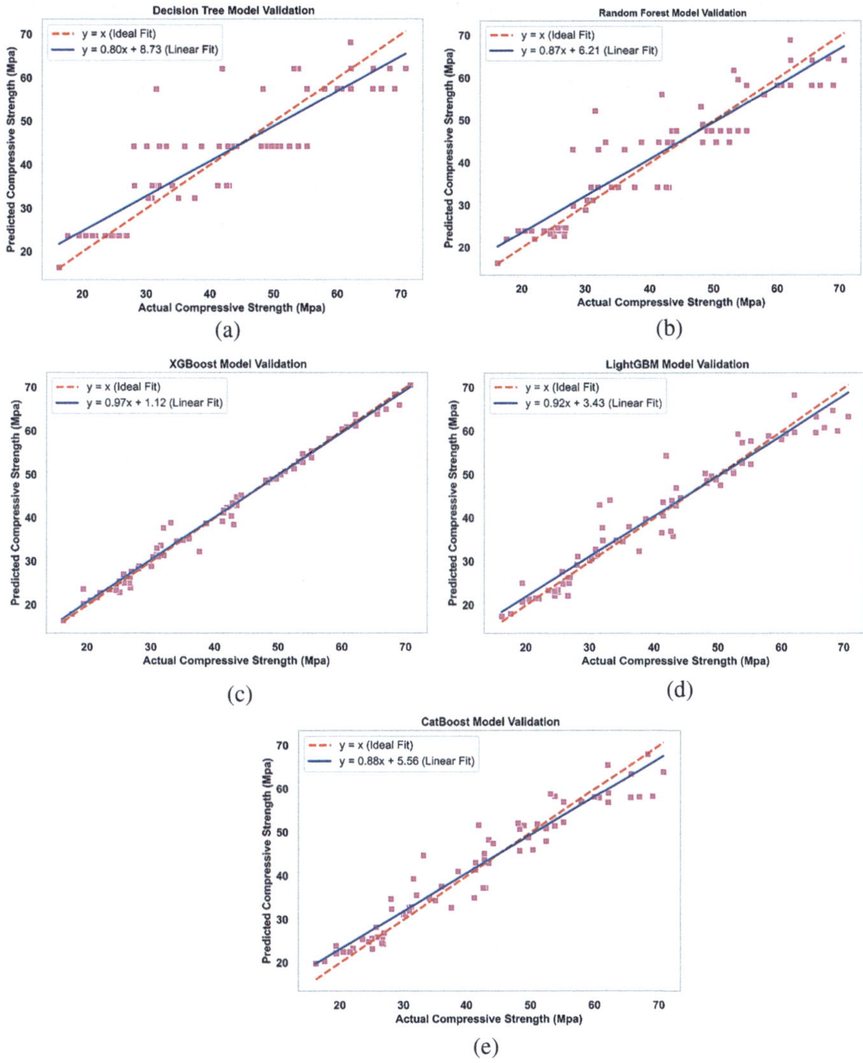

Fig. 1.11 Regression slope analysis of models (**a**. Decision Tree, **b**. Random Forest, **c**. XGBoost, **d**. LightGBM, **e**. CatBoost) on validation data

Table 1.3 Overview of model performance metrics

Data set	Algorithm	R^2	RMSE	MAE	RSR	a10-index	a20-index
Training	DT	0.90	4.38	6.14	0.43	56.60	88.07
	RF	0.94	3.65	4.92	0.35	59.72	89.35
	XGBoost	1.00	0.54	1.10	0.08	94.71	100.00
	LightGBM	0.97	2.48	3.26	0.23	79.63	94.44
	CatBoost	0.99	1.84	2.40	0.17	83.80	98.61
Testing	DT	0.90	4.62	6.65	0.45	54.43	87.97
	RF	0.94	4.54	5.59	0.36	40.74	90.74
	XGBoost	0.97	2.83	3.51	0.23	72.07	90.89
	LightGBM	0.96	3.33	4.15	0.27	68.52	88.89
	CatBoost	0.96	3.00	3.81	0.29	74.22	88.74
Validation	DT	0.87	5.46	7.10	0.49	43.04	83.54
	RF	0.96	3.80	5.09	0.32	51.85	92.59
	XGBoost	0.99	2.12	2.26	0.17	77.67	100.00
	LightGBM	0.99	2.70	3.13	0.19	77.78	96.30
	CatBoost	0.98	3.23	4.69	0.27	66.78	92.58

dataset. Furthermore, it consistently outperformed the other models across different performance measures, achieving a minimum MAE of 1.10 for testing and 3.51 for training, and the evaluation metrics are visualized in Fig. 1.12.

1.4.4 SHAP Interpretation

The significance ranking of input attributes in predicting the model's output was determined using average SHAP values, as depicted in Fig. 1.13. The mean of each feature's SHAP value over the whole dataset yields this number. The characteristics that are thought to have the most influence on output prediction are those having the largest absolute magnitudes in SHAP values. In predicting the compressive capacity of concrete, the water–cement ratio (W/C) holds the greatest influence, as reflected by its highest SHAP value.

SHAP-based importance illustrates how influential the input variables are in comparison to one another, yet it does not specify whether their impact on the predicted result is positive or negative. Figure 1.14 presents the SHAP summary visualization, which demonstrates how individual input attributes impact the predicted outcome.

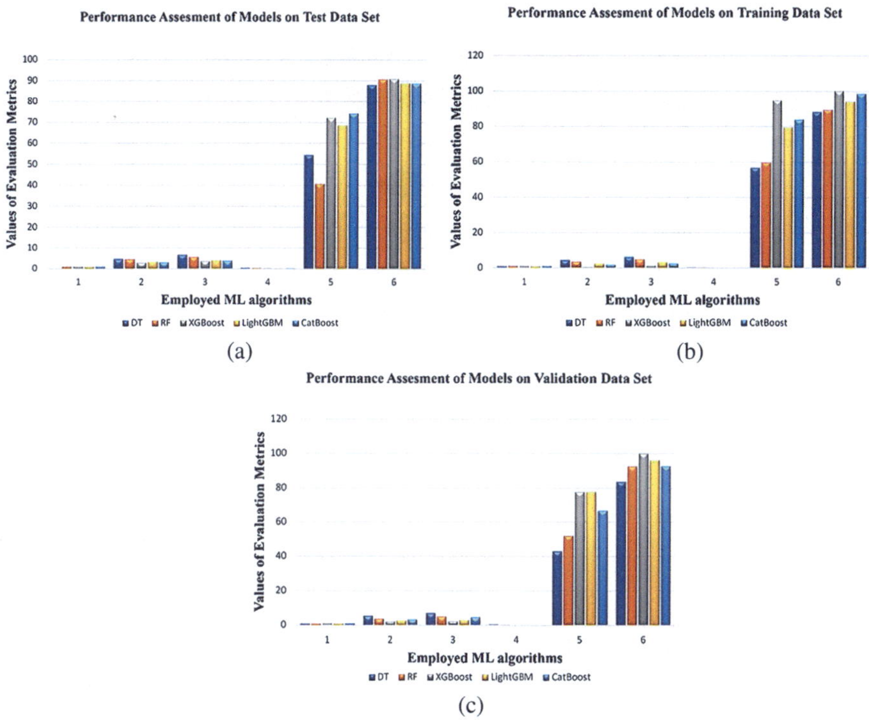

Fig. 1.12 Visualization of evaluation metrics: (**a**) on Test dataset, (**b**) on Training dataset, (**c**) on Validation dataset

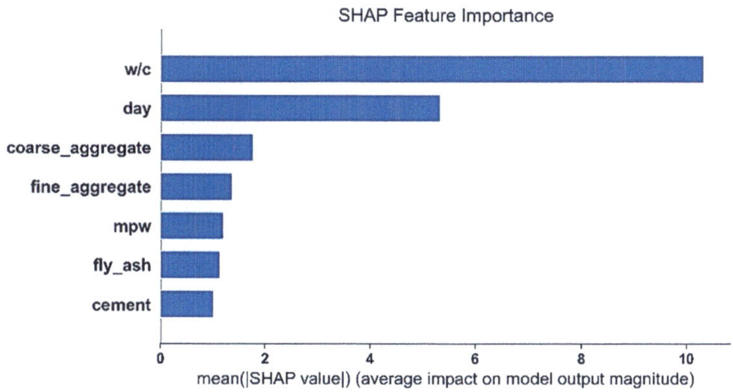

Fig. 1.13 Importance of SHAP global features

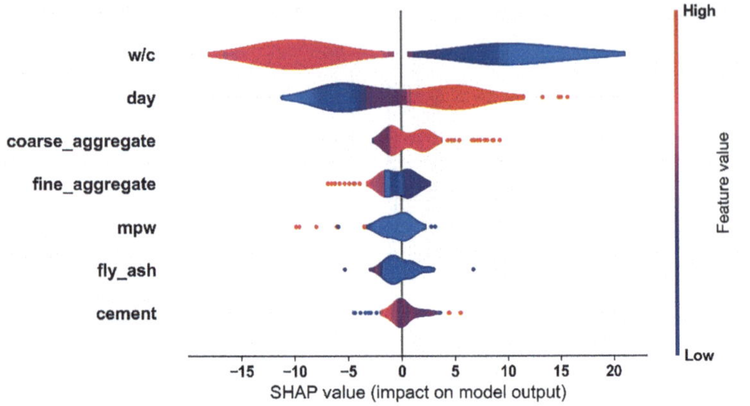

Fig. 1.14 SHAP summary plot

1.5 Conclusion

This study analyzed the predictive capability of several machine learning algorithms: Decision Tree (DT), Random Forest (RF), XGBoost, LightGBM, and CatBoost for estimating the compressive strength of concrete incorporating fly ash and MCSW. A collection of 270 experimental records gathered from published sources was employed, and the algorithms were developed and tested using tuned hyperparameters.

- Results revealed that all models exhibited good predictive ability, but XGBoost steadily exceeded the performance of the others, attaining the greatest accuracy with an R^2 of 0.97 on the test set with the smallest error values (MAE = 3.51, RMSE = 2.83).
- The SHAP-based feature analysis confirmed that the water–cement ratio (W/C) was the most influential variable; next in importance were curing duration and cement quantity.
- These findings confirm that ML approaches, particularly ensemble-based boosting algorithms, can provide fast, cost-effective, and reliable alternatives to experimental testing.

By enabling accurate strength prediction, this research supports the development of eco-friendly and sustainable concrete mix designs utilizing industrial waste, thus reducing cement consumption and lowering CO_2 emissions in the construction industry.

1.6 Recommendation

1. Future studies could explore hybrid models (e.g., combining ML with optimization algorithms or deep neural networks) to further improve predictive accuracy.
2. Beyond compressive strength, models should also predict durability indices (e.g., chloride resistance, permeability, carbonation depth) for a more holistic performance evaluation.
3. Extending the framework to include life cycle assessment (LCA), cost-benefit analysis, and environmental impact indicators will provide deeper insights into sustainable mix design.

1.7 Limitations

As the generalizability of ML model increases with the increase of dataset, in this study only 270 data points were used, if can be made better if more data set is used. To have more data, more experiments should be conducted using various mixed proportions of fly ash and MCSW.

AI Use Disclosure Statement The authors did not use any generative AI or AI-assisted technologies in the creation of this manuscript.

References

1. Liu G et al (2020) Cyber-physical system-based real-time monitoring and visualization of greenhouse gas emissions of prefabricated construction. J Clean Prod 246:119059. https://doi.org/10.1016/J.JCLEPRO.2019.119059
2. Hassan HS et al (2019) Cleaner production of one-part white geopolymer cement using pre-treated wood biomass ash and diatomite. J Clean Prod 209:1420–1428. https://doi.org/10.1016/J.JCLEPRO.2018.11.137
3. Akbarzadeh Bengar H, Shahmansouri AA (2020) A new anchorage system for CFRP strips in externally strengthened RC continuous beams. J Build Eng 30:101230. https://doi.org/10.1016/J.JOBE.2020.101230
4. Benhelal E, Shamsaei E, Rashid MI (2019) Novel modifications in a conventional clinker making process for sustainable cement production. J Clean Prod 221:389–397. https://doi.org/10.1016/J.JCLEPRO.2019.02.259
5. Samimi K, Kamali-Bernard S, Akbar Maghsoudi A, Maghsoudi M, Siad H (2017) Influence of pumice and zeolite on compressive strength, transport properties and resistance to chloride penetration of high strength self-compacting concretes. Constr Build Mater 151:292–311. https://doi.org/10.1016/J.CONBUILDMAT.2017.06.071
6. Taji I et al (2019) Application of statistical analysis to evaluate the corrosion resistance of steel rebars embedded in concrete with marble and granite waste dust. J Clean Prod 210:837–846. https://doi.org/10.1016/J.JCLEPRO.2018.11.091

7. Tang P, Chen W, Xuan D, Zuo Y, Poon CS (2020) Investigation of cementitious properties of different constituents in municipal solid waste incineration bottom ash as supplementary cementitious materials. J Clean Prod 258:120675. https://doi.org/10.1016/J.JCLEPRO.2020.120675
8. Elsayed M, Badawy S, Tayeh BA, Elymany M, Salem M, ElGawady M (2022) Shear behaviour of ultra-high performance concrete beams with openings. Structure 43:546–558. https://doi.org/10.1016/J.ISTRUC.2022.06.071
9. Amran M, Debbarma S, Ozbakkaloglu T (2021) Fly ash-based eco-friendly geopolymer concrete: a critical review of the long-term durability properties. Constr Build Mater 270:121857. https://doi.org/10.1016/J.CONBUILDMAT.2020.121857
10. Manjunatha M, Seth D, Balaji KVGD, Roy S, Tangadagi RB (2023) Utilization of industrial-based PVC waste powder in self-compacting concrete: a sustainable building material. J Clean Prod 428:139428. https://doi.org/10.1016/J.JCLEPRO.2023.139428
11. Elkerany AM, Keshta MM, Elshikh MMY, Elshami AA, Youssf O (2023) Characteristics of sustainable concrete containing metakaolin and magnetized water. Buildings 13(6):1430. https://doi.org/10.3390/BUILDINGS13061430
12. Keshta MM, Elshikh MMY, Abd Elrahman M, Youssf O (2022) Utilizing of magnetized water in enhancing of volcanic concrete characteristics. J Compos Sci 6(10):320. https://doi.org/10.3390/JCS6100320
13. Manjunatha M, Preethi S, Malingaraya, Mounika HG, Niveditha KN, Ravi (2021) Life cycle assessment (LCA) of concrete prepared with sustainable cement-based materials. Mater Today Proc 47:3637–3644. https://doi.org/10.1016/J.MATPR.2021.01.248
14. Albiajawi MI, Embong R, Muthusamy K (2022) An overview of the utilization and method for improving pozzolanic performance of agricultural and industrial wastes in concrete. Mater Today Proc 48:778–783. https://doi.org/10.1016/J.MATPR.2021.02.260
15. Sardinha M, de Brito J, Rodrigues R (2016) Durability properties of structural concrete containing very fine aggregates of marble sludge. Constr Build Mater 119:45–52. https://doi.org/10.1016/J.CONBUILDMAT.2016.05.071
16. Kavya C, Mohan UV (2022) Properties of SCC with marble powder as a marginal material. Mater Today Proc 60:2167–2170. https://doi.org/10.1016/J.MATPR.2022.02.403
17. Ghani A et al (2022) Properties of SCC with marble powder as a marginal material. Mater Today Proc 14(1). https://doi.org/10.1038/S41598-024-70303-Y
18. Meera M, Prashad BDV, Gupta S (2018) Experimental investigations on concrete with fly ash and marble powder for paver blocks. Int J Eng Technol (UAE) 7(3):120–122. https://doi.org/10.14419/IJET.V7I3.32.18410
19. Anandaraj S et al (2022) An experimental study on fly ash (FA) and marble powder in the properties of self-compacting concrete (SCC). Mater Today Proc 52:1771–1774. https://doi.org/10.1016/J.MATPR.2021.11.441
20. Meera M, Gupta S (2024) Performance evaluation of marble powder and fly ash concrete for non-structural applications. J Build Eng 84:108499. https://doi.org/10.1016/J.JOBE.2024.108499
21. Ghani A et al (2024) Experimental study on the mechanical behavior of concrete incorporating fly ash and marble powder waste. Sci Rep 14(1):1–18. https://doi.org/10.1038/s41598-024-70303-y
22. Singhal V, Nagar R, Agrawal V (2020) Sustainable use of fly ash and waste marble slurry powder in concrete. Mater Today Proc 32:975–981. https://doi.org/10.1016/J.MATPR.2020.05.650
23. Filali S, Nasser A (2024) Evaluating the impact of marble waste and fly ash as sand replacements on concrete's compressive strength and workability. Eng Technol Appl Sci Res 14(5):16797–16801. https://doi.org/10.48084/ETASR.8234
24. Chou JS, Tsai CF, Pham AD, Lu YH (2014) Machine learning in concrete strength simulations: multi-nation data analytics. Constr Build Mater 73:771–780. https://doi.org/10.1016/J.CONBUILDMAT.2014.09.054
25. Yeh IC (1998) Modeling of strength of high-performance concrete using artificial neural networks. Cem Concr Res 28(12):1797–1808. https://doi.org/10.1016/S0008-8846(98)00165-3

26. Ziolkowski P, Niedostatkiewicz M (2019) Machine learning techniques in concrete mix design. Materials 12(8):1256. https://doi.org/10.3390/MA12081256
27. Abuodeh OR, Abdalla JA, Hawileh RA (2020) Assessment of compressive strength of ultra-high performance concrete using deep machine learning techniques. Appl Soft Comput 95:106552. https://doi.org/10.1016/J.ASOC.2020.106552
28. Feng DC et al (2020) Machine learning-based compressive strength prediction for concrete: an adaptive boosting approach. Constr Build Mater 230:117000. https://doi.org/10.1016/J.CONBUILDMAT.2019.117000
29. Behnood A, Golafshani EM (2020) Machine learning study of the mechanical properties of concretes containing waste foundry sand. Constr Build Mater 243:118152. https://doi.org/10.1016/J.CONBUILDMAT.2020.118152
30. Kandiri A, Mohammadi Golafshani E, Behnood A (2020) Estimation of the compressive strength of concretes containing ground granulated blast furnace slag using hybridized multi-objective ANN and salp swarm algorithm. Constr Build Mater 248:118676. https://doi.org/10.1016/J.CONBUILDMAT.2020.118676
31. Salimbahrami SR, Shakeri R (2021) Experimental investigation and comparative machine-learning prediction of compressive strength of recycled aggregate concrete. Soft Comput 25(2):919–932. https://doi.org/10.1007/S00500-021-05571-1/METRICS
32. Shahmansouri AA, Akbarzadeh Bengar H, Jahani E (2019) Predicting compressive strength and electrical resistivity of eco-friendly concrete containing natural zeolite via GEP algorithm. Constr Build Mater 229:116883. https://doi.org/10.1016/J.CONBUILDMAT.2019.116883
33. Jijo BT, Abdulazeez AM (2021) Classification based on decision tree algorithm for machine learning. J Appl Sci Technol Trends 2(1):20–28. https://doi.org/10.38094/jastt20165
34. Su J, Zhang H (2025) A fast decision tree learning algorithm introduction and related work. www.aaai.org. Accessed 14 Aug 2025
35. Ben-Haim Y, Com YI (2010) A streaming parallel decision tree algorithm Elad Tom-Tov. J Mach Learn Res 11:849–872
36. Navada A, Ansari AN, Patil S, Sonkamble BA (2011) Overview of use of decision tree algorithms in machine learning. In: Proceedings—2011 IEEE control and system graduate research colloquium, ICSGRC 2011, pp 37–42. https://doi.org/10.1109/ICSGRC.2011.5991826
37. Breiman L (2001) Random forests. Mach Learn 45(1):5–32. https://doi.org/10.1023/A:1010933404324/METRICS
38. Livingston F (2005) Implementation of Breiman's random forest machine learning algorithm
39. Liu Y, Wang Y, Zhang J (2012) New machine learning algorithm: random forest. In: Information computing and applications. ICICA 2012, Lecture notes in computer science (including subseries lecture notes in artificial intelligence and lecture notes in bioinformatics), vol 7473, pp 246–252. https://doi.org/10.1007/978-3-642-34062-8_32
40. Jaiswal JK, Samikannu R (2017) Application of random forest algorithm on feature subset selection and classification and regression. In: Proceedings of the 2nd world congress on computing and communication technologies, WCCCT 2017, pp 65–68. https://doi.org/10.1109/WCCCT.2016.25
41. Kavzoglu T, Teke A (2022) Advanced hyperparameter optimization for improved spatial prediction of shallow landslides using extreme gradient boosting (XGBoost). Bull Eng Geol Environ 81(5):1–22. https://doi.org/10.1007/S10064-022-02708-W/METRICS
42. Chen T, Guestrin C (2016) XGBoost: a scalable tree boosting system. In: Proceedings of the ACM SIGKDD international conference on knowledge discovery and data mining, 13–17 Aug 2016, pp 785–794. https://doi.org/10.1145/2939672.2939785
43. Bakouregui AS, Mohamed HM, Yahia A, Benmokrane B (2021) Explainable extreme gradient boosting tree-based prediction of load-carrying capacity of FRP-RC columns. Eng Struct 245:112836. https://doi.org/10.1016/J.ENGSTRUCT.2021.112836
44. Hengl T et al (2017) Soil nutrient maps of sub-Saharan Africa: assessment of soil nutrient content at 250 m spatial resolution using machine learning. Nutr Cycl Agroecosyst 109(1):77–102. https://doi.org/10.1007/S10705-017-9870-X

45. Zhang D, Qian L, Mao B, Huang C, Huang B, Si Y (2018) A data-driven design for fault detection of wind turbines using random forests and XGboost. IEEE Access 6:21020–21031. https://doi.org/10.1109/ACCESS.2018.2818678
46. Liang W, Luo S, Zhao G, Wu H (2020) Predicting hard rock pillar stability using GBDT, XGBoost, and LightGBM algorithms. Mathematics 8(5):765. https://doi.org/10.3390/MATH8050765
47. Zeng H et al (2019) A LightGBM-based EEG analysis method for driver mental states classification. Comput Intell Neurosci 2019. https://doi.org/10.1155/2019/3761203
48. Abdulalim Alabdullah A, Iqbal M, Zahid M, Khan K, Nasir Amin M, Jalal FE (2022) Prediction of rapid chloride penetration resistance of metakaolin based high strength concrete using light GBM and XGBoost models by incorporating SHAP analysis. Constr Build Mater 345:128296. https://doi.org/10.1016/J.CONBUILDMAT.2022.128296
49. Khan M, Javed MF, Othman NA, Ur Rehman SK, Ahmad F (2025) Predicting penetration depth in ultra-high-performance concrete targets under ballistic impact: an interpretable machine learning approach augmented by deep generative adversarial network. Results Eng 25:103909. https://doi.org/10.1016/J.RINENG.2024.103909
50. Sapkota SC, Saha P, Das S, Meesaraganda LVP (2024) Prediction of the compressive strength of normal concrete using ensemble machine learning approach. Asian J Civil Eng 25(1):583–596. https://doi.org/10.1007/S42107-023-00796-X/METRICS
51. Wang X, Banthia N, Yoo DY (2024) Reinforcement bond performance in 3D concrete printing: explainable ensemble learning augmented by deep generative adversarial networks. Autom Constr 158:105164. https://doi.org/10.1016/J.AUTCON.2023.105164
52. Partial replacement of cement with marble dust and fly ash. www.internationaljournalssrg.org. Accessed 16 Aug 2025
53. Rid ZA, Shah SNR, Memon MJ, Jhatial AA, Keerio MA, Goh WI (2022) Evaluation of combined utilization of marble dust powder and fly ash on the properties and sustainability of high-strength concrete. Environ Sci Pollut Res 29(19):28005–28019. https://doi.org/10.1007/S11356-021-18379-1/METRICS
54. Khan EA et al (2023) Utilization of waste marble powder as partial replacement of cement in engineered cementitious composite. Cogent Eng 10(1). https://doi.org/10.1080/23311916.2023.2243749
55. Ghani A, Ali Z, Khan FA, Shah SR, Khan SW, Rashid M (2020) Experimental study on the behavior of waste marble powder as partial replacement of sand in concrete. SN Appl Sci 2(9):1–13. https://doi.org/10.1007/S42452-020-03349-Y/FIGURES/15
56. Singh M, Choudhary K, Srivastava A, Singh Sangwan K, Bhunia D (2017) A study on environmental and economic impacts of using waste marble powder in concrete. J Build Eng 13:87–95. https://doi.org/10.1016/J.JOBE.2017.07.009
57. Abbas MM, Muntean R (2025) Marble powder as a sustainable cement replacement: a review of mechanical properties. Sustainability 17(2):736. https://doi.org/10.3390/SU17020736
58. Ashish DK, Verma SK, Kumar R, Sharma N (2016) Properties of concrete incorporating sand and cement with waste marble powder. Adv Concr Constr 4(2):145–160. https://doi.org/10.12989/ACC.2016.4.2.145
59. Kumar A, Rohit Kumar M (2018) A review paper based on partial replacement of cement with marble powder and fly ash for M25 grade concrete. Int J Technol Res Eng 5(10). www.ijtre.com. Accessed 16 Aug 2025
60. Hamza M, Shahzada K, Iqbal M, Khan EA (2024) Evaluation of mechanical properties of paver blocks by replacing cement with fly ash and marble waste. Discov Appl Sci 6(5):1–20. https://doi.org/10.1007/S42452-024-05843-Z/TABLES/14
61. Neville AM (1995) Properties of concrete, 4th and final ed. Prentice Hall, p 844
62. Shehadeh A, Alshboul O, Al Mamlook RE, Hamedat O (2021) Machine learning models for predicting the residual value of heavy construction equipment: an evaluation of modified decision tree, LightGBM, and XGBoost regression. Autom Constr 129:103827. https://doi.org/10.1016/J.AUTCON.2021.103827

63. Yang S, Zhang H, Yang S, Zhang H (2018) Comparison of several data mining methods in credit card default prediction. Intell Inf Manag 10(5):115–122. https://doi.org/10.4236/IIM.2018.105010
64. Gong X, Zheng B, Xu G, Chen H, Chen C (2021) Application of machine learning approaches to predict the 5-year survival status of patients with esophageal cancer. J Thorac Dis 13(11):6240–6251. https://doi.org/10.21037/JTD-21-1107
65. Ma L, Zhou C, Lee D, Zhang J (2022) Prediction of axial compressive capacity of CFRP-confined concrete-filled steel tubular short columns based on XGBoost algorithm. Eng Struct 260:114239. https://doi.org/10.1016/J.ENGSTRUCT.2022.114239
66. Leevy JL, Hancock J, Zuech R, Khoshgoftaar TM (2020) Detecting cybersecurity attacks using different network features with LightGBM and XGBoost learners. In: Proceedings of the 2020 IEEE 2nd international conference on cognitive machine intelligence, CogMI 2020, pp 190–197. https://doi.org/10.1109/COGMI50398.2020.00032
67. Liang M, Chang Z, Wan Z, Gan Y, Schlangen E, Šavija B (2022) Interpretable ensemble-machine-learning models for predicting creep behavior of concrete. Cem Concr Compos 125:104295. https://doi.org/10.1016/J.CEMCONCOMP.2021.104295
68. Amin MN, Khan K, Abu Arab AM, Farooq F, Eldin SM, Javed MF (2023) Prediction of sustainable concrete utilizing rice husk ash (RHA) as supplementary cementitious material (SCM): optimization and hyper-tuning. J Mater Res Technol 25:1495–1536. https://doi.org/10.1016/J.JMRT.2023.06.006
69. Yang D et al (2023) Compressive strength prediction of concrete blended with carbon nanotubes using gene expression programming and random forest: hyper-tuning and optimization. J Mater Res Technol 24:7198–7218. https://doi.org/10.1016/J.JMRT.2023.04.250
70. He S, Wu J, Wang D, He X (2022) Predictive modeling of groundwater nitrate pollution and evaluating its main impact factors using random forest. Chemosphere 290:133388. https://doi.org/10.1016/J.CHEMOSPHERE.2021.133388
71. Valatsos P, Vafeiadis T, Nizamis A, Ioannidis D, Tzovaras D (2021) Freight transportation route time prediction with ensemble learning techniques. In: ACM international conference proceeding series, pp 52–57. https://doi.org/10.1145/3503823.3503833
72. Ahmad GN, Fatima H, Ullah S, Saidi AS, Imdadullah (2022) Efficient medical diagnosis of human heart diseases using machine learning techniques with and without GridSearchCV. IEEE Access 10:80151–80173. https://doi.org/10.1109/ACCESS.2022.3165792
73. Dong W, Huang Y, Lehane B, Ma G (2020) XGBoost algorithm-based prediction of concrete electrical resistivity for structural health monitoring. Autom Constr 114:103155. https://doi.org/10.1016/J.AUTCON.2020.103155
74. Mahmood J, Mustafa GE, Ali M (2022) Accurate estimation of tool wear levels during milling, drilling and turning operations by designing novel hyperparameter tuned models based on LightGBM and stacking. Measurement 190:110722. https://doi.org/10.1016/J.MEASUREMENT.2022.110722
75. Zhu Y, Ahmad A, Ahmad W, Vatin NI, Mohamed AM, Fathi D (2022) Predicting the splitting tensile strength of recycled aggregate concrete using individual and ensemble machine learning approaches. Crystals (Basel) 12(5):569. https://doi.org/10.3390/CRYST12050569/S1
76. Obianyo II et al (2024) Machine learning models for predicting the compressive strength of agro-waste stabilized bricks for sustainable buildings. Discov Civil Eng 1(1):1–20. https://doi.org/10.1007/S44290-024-00052-4
77. Fei Z, Liang S, Cai Y, Shen Y (2023) Ensemble machine-learning-based prediction models for the compressive strength of recycled powder mortar. Materials 16(2):583. https://doi.org/10.3390/MA16020583
78. Zhang Y, Ren W, Chen Y, Mi Y, Lei J, Sun L (2024) Predicting the compressive strength of high-performance concrete using an interpretable machine learning model. Sci Rep 14(1):1–18. https://doi.org/10.1038/S41598-024-79502-Z

79. Peng F, Li Y, Xue W (2024) Transfer learning-based confinement strength prediction of concrete confined by FRP transverse reinforcements. Eng Struct 310:118116. https://doi.org/10.1016/J.ENGSTRUCT.2024.118116
80. Yaseen ZM et al (2018) Predicting compressive strength of lightweight foamed concrete using extreme learning machine model. Adv Eng Softw 115:112–125. https://doi.org/10.1016/J.ADVENGSOFT.2017.09.004
81. Yucel M, Namli E (2022) High performance concrete (HPC) compressive strength prediction with advanced machine learning methods: combinations of machine learning algorithms with bagging, rotation forest, and additive regression. In: Research anthology on machine learning techniques, methods, and applications, pp 68–89. https://doi.org/10.4018/978-1-6684-6291-1.CH006
82. Moriasi DN, Arnold JG, Van Liew MW, Bingner RL, Harmel RD, Veith TL (2007) Model evaluation guidelines for systematic quantification of accuracy in watershed simulations. Trans ASABE 50(3):885–900. https://doi.org/10.13031/2013.23153
83. Dabiri H, Kioumarsi M, Kheyroddin A, Kandiri A, Sartipi F (2022) Compressive strength of concrete with recycled aggregate; a machine learning-based evaluation. Clean Mater 3. https://doi.org/10.1016/J.CLEMA.2022.100044
84. Ahmad A, Ostrowski KA, Maślak M, Farooq F, Mehmood I, Nafees A (2021) Comparative study of supervised machine learning algorithms for predicting the compressive strength of concrete at high temperature. Materials 14(15). https://doi.org/10.3390/MA14154222
85. Chen Z, Xiao F, Guo F, Yan J (2023) Interpretable machine learning for building energy management: a state-of-the-art review. Adv Appl Energy 9:100123. https://doi.org/10.1016/J.ADAPEN.2023.100123
86. Zhao L (2020) Event prediction in big data era: a systematic survey. arXiv preprint arXiv:2007.09815. https://par.nsf.gov/servlets/purl/10213676. Accessed 17 Aug 2025
87. Iqbal MF et al (2020) Prediction of mechanical properties of green concrete incorporating waste foundry sand based on gene expression programming. J Hazard Mater 384:121322. https://doi.org/10.1016/J.JHAZMAT.2019.121322
88. Farooq F et al (2020) A comparative study of random forest and genetic engineering programming for the prediction of compressive strength of high strength concrete (HSC). Appl Sci 10(20):7330. https://doi.org/10.3390/APP10207330

Chapter 2
Interpretable Machine Learning for Public Bus Service Efficiency: A SHAP-Driven Framework for Operational Analytics

Mahbub Hassan, Md Kamrul Islam, Md Ehtesamul Haque, and Md Ashequl Islam

Abstract Public bus transport systems in emerging economies face persistent challenges in achieving operational efficiency, service coverage, and financial sustainability. This study presents an integrated analytical framework that combines predictive modeling, explainable artificial intelligence (XAI), and unsupervised clustering to diagnose and optimize fleet operations, with application to the Andhra Pradesh State Road Transport Corporation (APSRTC) in India. High-resolution planned and monitored operational datasets were used to train Light Gradient Boosting Machine (LightGBM), Random Forest, and Ridge Regression models for predicting fuel consumption. Model interpretability was achieved through SHapley Additive exPlanations (SHAP) to measure the relative contribution of predictor variables, while k-means clustering in a principal component analysis (PCA) reduced feature space was applied to identify service groups with distinct operational characteristics. The best performing models produced root mean squared error (RMSE) values between 18.18 and 20.48 L and coefficient of determination (R^2) values up to 0.71. Trip distance consistently appeared as the most influential predictor in both planned and monitored datasets. Clustering analysis identified three operational groups: long-haul high consumption, short-haul low consumption,

M. Hassan
Department of Civil Engineering, Faculty of Engineering, Chulalongkorn University, Bangkok, Thailand

M. K. Islam (✉)
Department of Civil and Environmental Engineering, College of Engineering, King Faisal University, Al Ahsa, Saudi Arabia
e-mail: maislam@kfu.edu.sa

M. E. Haque
Department of Computer Science, College of Computer Science and Information Technology (CCSIT), King Faisal University, Al Ahsa, Saudi Arabia
e-mail: mehaque@kfu.edu.sa

M. A. Islam
School of Electrical and Mechanical Engineering, College of Engineering and Information Technology, Adelaide University, Adelaide, Australia

M. Mirrashid et al. (eds.), *Intelligent Predictive Systems*, Emerging Trends in Mechatronics, https://doi.org/10.1007/978-981-95-7766-8_2

and intermediate services, providing a basis for targeted operational improvement. The integration of predictive modeling, SHAP-based interpretability, and operational segmentation enables two levels of insight: fleet-level performance diagnosis and route-specific strategic analysis. The findings support data-driven recommendations for improving fuel efficiency, reducing operational costs, and optimizing service planning. The proposed framework offers a replicable approach for state-owned transport corporations operating in similar contexts.

Keywords Public transport analytics · Explainable artificial intelligence (XAI) · SHAP interpretability · Machine learning in transportation · Operational efficiency optimization

2.1 Introduction

Public bus transport systems are a critical component of sustainable urban and intercity mobility, particularly in emerging economies where affordable, reliable, and accessible mass transit is essential for socioeconomic inclusion [1–3]. Beyond their social role, bus systems help reduce traffic congestion, lower per capita greenhouse gas (GHG) emissions, and improve energy efficiency compared to private vehicles. According to the International Energy Agency (IEA 2022) [4], efficiency improvements in the public transport sector can significantly accelerate progress toward national climate commitments.

In India, state-owned transport corporations such as the Andhra Pradesh State Road Transport Corporation (APSRTC) operate extensive fleets that cover large geographical areas [5]. Managing such large-scale operations requires balancing service coverage, operational efficiency, and financial sustainability. However, these systems face persistent challenges including fluctuating passenger demand, variable route conditions, rising fuel prices, and inefficiencies that arise from the gap between planned targets and monitored performance [6].

Fuel expenditure often accounts for 30–40% of total operating costs in state-owned bus fleets [7, 8], making fuel efficiency a key determinant of economic sustainability. Even small inefficiencies in route planning, vehicle allocation, or load management can result in substantial financial and environmental impacts at the fleet level [9, 10]. Although transport agencies routinely prepare planned schedules and performance targets, the actual monitored data often differ considerably from these projections due to unpredictable factors such as traffic conditions, vehicle breakdowns, and irregular passenger patterns [11].

Traditional operational analysis methods typically rely on aggregated key performance indicators (KPIs), which can obscure the complexity of operational dynamics. This limitation reduces the ability of planners to design targeted interventions for cost savings, service improvement, and environmental gains. To address these limitations, there is an increasing need for data-driven, interpretable, and actionable analytical frameworks that can operate effectively on real-world operational datasets [12].

The digitization of fleet operations through ticketing systems, GPS tracking, and fuel monitoring has made it possible to collect high-resolution transport data [12]. This advancement has opened new opportunities for machine learning (ML) in predicting key operational outcomes such as fuel consumption, occupancy rates, and revenue generation [13–15]. However, although high-performing models can be developed using algorithms such as Light Gradient Boosting Machine (LightGBM) [16, 17], their opaque nature often limits interpretability, which is crucial in public sector decision-making where transparency, accountability, and stakeholder trust are essential.

Explainable AI (XAI) methods, particularly SHapley Additive exPlanations (SHAP) [18, 19], address this issue by quantifying the contribution of each input variable to the model output [20]. SHAP provides a consistent, model-independent interpretability framework that helps identify operational levers, allowing transport managers to understand not only what is happening but also why it is happening.

While predictive models and interpretability tools reveal performance drivers, it is also important for transport managers to recognize that different services or routes may exhibit distinct operational patterns. Unsupervised learning techniques such as k-means clustering [21, 22] can group services into clusters with similar characteristics. When applied within a principal component analysis (PCA) reduced feature space [23], clustering offers a visual and quantitative segmentation of services, helping planners identify which routes are fuel-intensive, high-revenue, low-occupancy, or operationally anomalous.

Combining predictive modeling, SHAP interpretability, and clustering within a unified analytical workflow enables a dual perspective: a global understanding of fleet-wide performance drivers and a local diagnostic view for targeted interventions.

Prior studies in public transport analytics have examined specific aspects such as route optimization [24], schedule design [25], and predictive maintenance [26]. However, few studies have integrated predictive modeling, explainable AI, and unsupervised clustering into a single framework applied to both planned and monitored operational datasets. This integration remains rare in large state-owned bus corporations in developing countries, where operational realities often differ markedly from planned targets.

This study introduces an integrated analytical framework to diagnose and optimize public bus operations using APSRTC operational data. The main contributions are as follows:

- **Predictive Modeling**: Development of LightGBM-based regression models for predicting fuel consumption and fuel efficiency using both planned and monitored datasets.

- **Model Interpretability**: Application of SHAP to quantify and visualize the influence of operational, route, and vehicle variables on model predictions, providing transparent insights for decision-making.
- **Operational Segmentation**: Implementation of k-means clustering in PCA-reduced feature space to classify services into operationally distinct clusters that support targeted management actions.
- **Comparative Analysis**: Systematic evaluation of the differences between planned and monitored data to identify performance gaps and their operational determinants.
- **Actionable Recommendations**: Translation of analytical results into evidence-based strategies for enhancing fuel efficiency, revenue generation, and service planning.

The remainder of this paper is organized as follows: Sect. 2.2 describes the dataset, preprocessing steps, and modeling methodology. Section 2.3 presents the results, including predictive performance, SHAP interpretability analyses, and cluster profiles. Section 2.4 discusses operational implications, limitations, and policy relevance. Finally, Sect. 2.5 concludes the paper and outlines directions for future research.

2.2 Methodology

This section describes the complete analytical pipeline used to diagnose and explain fuel consumption in APSRTC bus operations. We work with a single tabular dataset of 1000 trip-level records and construct two feature regimes from the same table: a *planning* regime that uses only variables available before dispatch, and a *monitoring* regime that augments planning variables with information observed after service completion. Predictive models are trained and compared across both regimes, the most accurate model is interpreted using SHAP, and services are segmented through unsupervised clustering in a principal component space. All modeling and evaluation procedures were designed to ensure reproducibility, interpretability, and statistical validity.

2.2.1 Dataset and Variables

The dataset contains 1000 operational records; each row corresponds to one bus trip. Variables include identifiers, route and vehicle descriptors, operational and financial performance, and temporal context. Table 2.1 presents the variables used in the analysis. The main supervised target is fuel consumed liters (trip level fuel use in liters).

Table 2.1 APSRTC variables used in this study

Variable	Type/unit	Description
Bus id	Categorical	Unique vehicle identifier
Route	Categorical	Origin to destination label
Bus type	Categorical	Service class (e.g., Volvo AC, Sleeper)
Depot	Categorical	Operating depot
Date	Date (YYYY MM DD)	Trip date used to derive month and day of week
Capacity	Integer (seats)	Seating capacity
Passengers	Integer (persons)	Passengers carried
Occupancy rate	Numeric (%)	Load factor (100 × passengers/capacity)
Distance (km)	Numeric (km)	Trip distance
Fare per passenger	Numeric (INR)	Mean fare per passenger
Revenue	Numeric (INR)	Total fare revenue
Fuel consumed liters	Numeric (L)	Fuel used (supervised target) month
	Categorical	Calendar month (derived) day of week
	Categorical	Day of week (derived)

Data validation confirmed the consistency of the financial variables. In particular, revenue satisfied the near identity in Eq. (2.1) up to rounding error; therefore, revenue is treated as a derived variable when discussing multicollinearity:

$$\text{Revenue} \approx \text{passengers} \times \text{fare per passenger}. \tag{2.1}$$

2.2.2 *Feature Regimes: Planning and Monitoring*

Two disjoint feature regimes are defined to represent the information available before dispatch and after trip completion.

2.2.2.1 Planning (PLAN)

The planning regime uses only ex ante variables (Eq. 2.2), which are appropriate for proactive budgeting and scheduling:

$$X_{\text{PLAN}} = \{r, b, d, m, w, c, \delta, f\}, \tag{2.2}$$

where r: route, b: bus type, d: depot, m: month, w: day of week, c: capacity, δ: distance (km), f: fare per passenger.

2.2.2.2 Monitoring (MONITOR)

The monitoring regime adds post-service utilization indicators (Eq. 2.3):

$$X_{\text{MON}} = X_{\text{PLAN}} \cup \{p, \rho\}, \tag{2.3}$$

where p: passengers, ρ: occupancy rate.

2.2.3 *Preprocessing and Encoders*

Dates are converted to the date time format and expanded into month and day of week. Categorical fields are transformed by one hot encoding. Let $x \in R^p$ represent the raw feature vector with categorical part $x^{(c)}$ and numerical part $x^{(n)}$. The column transformer applies the mapping in Eq. (2.4):

$$\Phi(x) = \left[\text{OHE}\left(x^{(c)}\right), \text{Std}\left(x^{(n)}\right)\right], \tag{2.4}$$

where OHE converts each categorical feature into binary indicators and $\text{Std}(x_j) = (x_j - \mu_j)/\sigma_j$ standardizes numerical features when required. Scaling is not needed for tree ensembles but is applied before PCA and clustering to ensure a well-conditioned covariance structure. No missing values were present. Outlier screening used robust z scores, and observations with $|z| > 3$ were checked against domain constraints.

2.2.4 *Supervised Learning Objectives and Losses*

Let y = fuel consumed liters $\in R$ and $f(\cdot)$ denote the regression function trained on $\Phi(X)$. For completeness, preliminary results are also computed for fuel efficiency $y^{\star}$ = distance km/fuel consumed liters although this second target is not used for interpretation or clustering.

Predictive accuracy is measured using Root Mean Squared Error (RMSE, Eq. 2.5), Mean Absolute Error (MAE, Eq. 2.6), and the coefficient of determination (R^2, Eq. 2.7):

$$\text{RMSE} = \sqrt{\frac{1}{n}\sum_{i=1}^{n}\left(y_i - \hat{y}_i\right)^2}, \tag{2.5}$$

$$\text{MAE} = \frac{1}{n}\sum_{i=1}^{n}\left|y_i - \hat{y}_i\right|, \tag{2.6}$$

$$R^2 = 1 - \frac{\sum_i (y_i - \hat{y}_i)^2}{\sum_i (y_i - \bar{y}_i)^2}. \tag{2.7}$$

These metrics collectively represent scale-sensitive error, absolute deviation, and explained variance.

2.2.5 Models and Training Protocol

The primary learner is LightGBM [27], a gradient-boosted tree ensemble that minimizes a second-order Taylor approximation of an additive objective. At iteration t, LightGBM fits a tree $T^{(t)}$ to the gradients and Hessians, choosing leaves that maximize the regularized gain (Eq. 2.8):

$$G = \sum_l \frac{\left(\sum_{i \in l} g_i\right)^2}{\sum_{i \in l} h_i + \lambda} - \gamma L, \tag{2.8}$$

where λ controls L_2 regularization on leaf weights and γ penalizes the number of leaves L. The ensemble prediction updates iteratively as $\hat{y}^{(t)} = \hat{y}^{(t-1)} + \eta T^{(t)}(\Phi(x))$ with learning rate η.

For context, Ridge regression and Random Forest baselines are also trained using the same folds and features. The dataset is split into training (80%) and hold-out (20%) subsets, with model selection conducted only on the training data. Generalization error is estimated through a repeated K-fold protocol (5 folds, 3 repeats, 15 folds total). For LightGBM, hyperparameters are tuned via Bayesian optimization within cross-validation. Each model's performance vector $\{m_1, \ldots, m_{15}\}$ is summarized by its mean and a nonparametric 95% confidence interval estimated by bootstrap resampling ($B = 2000$ replicates).

2.2.6 SHAP Explanations for Tree Ensembles

Model transparency is achieved using SHapley Additive exPlanations (SHAP) [28]. For an instance x, SHAP provides the additive decomposition (Eq. 2.9):

$$f(\Phi(x)) - E\left[f(\Phi(X))\right] = \sum_{j=1}^{p} \varphi_j(x), \tag{2.9}$$

where $\varphi_j(x)$ represents the Shapley value of feature j, calculated as the game theoretic average of marginal contributions across all feature subsets. The tree-specific explainer uses the structure of decision trees for efficient exact or near-exact computation. Global importance is evaluated by the expectation of absolute contributions $E[|\varphi_j|]$, and local effects are explored through dependence plots that display nonlinear and interaction patterns. SHAP is applied to the most accurate fuel consumption model in each regime (PLAN and MONITOR) to identify both pre-dispatch and post-service operational levers.

2.2.7 *Unsupervised Segmentation in Principal Component Space*

To explore operational heterogeneity, clustering is applied to a compact representation of the monitored feature space. Let $Z \in R^{n \times d}$ denote the standardized matrix of selected operational variables. Principal Component Analysis (PCA) projects Z onto uncorrelated axes through the eigendecomposition of the covariance matrix (Eq. 2.10):

$$S = \frac{1}{n-1} Z^T Z = V \Lambda V^T, \quad U = ZV, \tag{2.10}$$

where U contains the PCA scores. The first two components are retained because they capture the main variance structure and provide an effective two-dimensional embedding for visual analysis.

Clustering uses k-means on the PCA scores. Given centroids $\{\mu_k\}_{k=1}^{K}$ assignments $\{C_k\}$, k-means minimizes the within-cluster distortion (Eq. 2.11):

$$J(\{\mu_k k\}, \{C_k\}) = \sum_{k=1}^{K} \sum_{i \in C_k} U_i - \mu_{k2}^{2} \tag{2.11}$$

Candidate values of K are evaluated using the silhouette index (Eq. 2.12):

$$s(i) = \frac{b(i) - a(i)}{\max\{a(i), b(i)\}} \tag{2.12}$$

where $a(i)$ is the average distance within a cluster and $b(i)$ the smallest average distance to another cluster. The optimal K maximizes the mean $s(i)$ while preserving interpretability.

2.2.8 *Statistical Comparison of Regimes*

To evaluate whether planning features are sufficient relative to monitoring features, we compare paired fold-level errors from the MONITOR and PLAN regimes using the Wilcoxon signed-rank test.

Let

$$\left\{e_j^{\text{MON}}\right\} \text{ and } \left\{e_j^{\text{PLAN}}\right\}$$

denote the RMSE values from each cross-validation fold. We compute the differences

$$d_j = e_j^{\text{MON}} - e_j^{\text{PLAN}},$$

rank the absolute differences $|d_j|$ and obtain a two-sided p-value under the null hypothesis that the median difference is zero. This nonparametric test is appropriate for non-Gaussian and heteroscedastic error distributions and respects the paired structure inherent to cross-validation.

2.2.9 Implementation and Reproducibility

All experiments were performed in Python. Preprocessing, cross-validation, and clustering were implemented with scikit-learn. Gradient boosting used LightGBM, and SHAP explanations used SHAP. Random seeds were fixed to ensure reproducibility.

2.3 Results

2.3.1 Predictive Performance on Fuel Consumption

Table 2.2 presents the cross-validated performance of the three learners under the two feature regimes. For both PLAN and MONITOR, *Ridge* achieved the lowest RMSE and MAE with the highest R^2 value, followed closely by Random Forest, while LightGBM performed slightly lower by approximately 1.5–2.3 L RMSE on average. The difference between PLAN and MONITOR regimes was small. A paired

Table 2.2 Cross-validated performance (mean with 95% CI over 15 folds) for fuel consumed liters. Bold values indicate the best model within each regime

Regime	Model	RMSE	MAE	R^2
MONITOR	**Ridge**	**18.18** [17.20, 19.38]	**12.87** [12.41, 13.35]	**0.710** [0.689, 0.728]
MONITOR	Random Forest	18.81 [17.69, 20.01]	13.23 [12.82, 13.64]	0.689 [0.667, 0.713]
MONITOR	LightGBM	20.48 [19.50, 21.61]	14.71 [14.30, 15.19]	0.632 [0.614, 0.649]
PLAN	**Ridge**	**18.17** [17.18, 19.37]	**12.85** [12.40, 13.32]	**0.710** [0.691, 0.729]
PLAN	Random Forest	18.79 [17.70, 20.02]	13.18 [12.72, 13.66]	0.690 [0.667, 0.711]
PLAN	LightGBM	20.38 [19.40, 21.43]	14.60 [14.18, 15.01]	0.635 [0.617, 0.654]

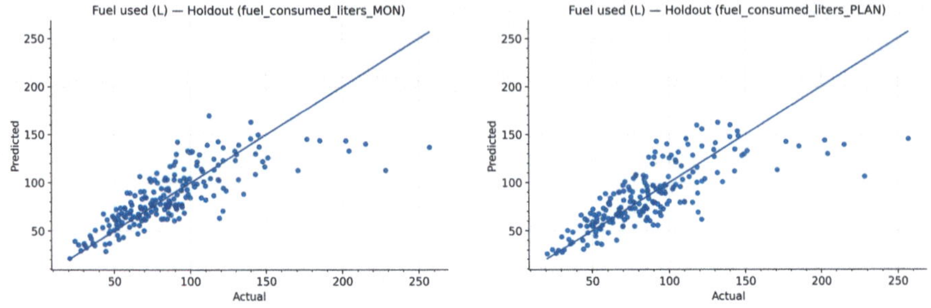

(a) MONITOR holdout (b) PLAN holdout

Fig. 2.1 Prediction versus actual fuel consumption (liters) on the held-out test set for (**a**) MONITOR and (**b**) PLAN regimes. Each panel displays model predictions plotted against observed fuel consumption with the 45° reference line indicating perfect agreement

nonparametric comparison of per fold errors for LightGBM across regimes produced a nonsignificant result (Wilcoxon signed-rank, $p = 0.6387$, $n = 15$ folds), indicating no consistent performance advantage of MONITOR over PLAN within this dataset.

Generalization on the held-out test set was consistent with cross-validation. The selected tree ensemble under the MONITOR regime reached RMSE = 23.05 L, MAE = 15.43 L, and $R^2 = 0.618$ (Fig. 2.1). Prediction versus actual scatter plots for both regimes cluster tightly around the 45° reference line, with slightly wider dispersion at higher volumes (150–200 L), indicating increased variance for long-distance services.

2.3.2 Explainability of Fuel Consumption with SHAP

To convert predictive accuracy into operational understanding, the LightGBM models were analyzed using SHAP. The global beeswarm plots (Fig. 2.2) show that *distance km* dominates the feature attribution spectrum in both regimes, followed by *fare per passenger* and *capacity* in PLAN, and by *occupancy rate* and *passengers* in MONITOR. The latter two are unavailable to planners before dispatch. Categorical features such as *route* and *bus type* contribute moderately, reflecting route geometry and vehicle class effects.

The local dependence plots (Fig. 2.3) show a monotonic and nearly linear increase of SHAP values with *distance km* across both regimes, which aligns with physical intuition: longer trips require more fuel. The slope becomes steeper for distances above approximately 350 km, indicating nonlinear effects such as extended highway operation or auxiliary energy demands. For *occupancy rate* (MONITOR only), SHAP values are slightly positive across most of the range, showing that higher load factors correspond to greater total fuel use at the trip level although per kilometer or per passenger intensity may decrease. The *capacity* plot in PLAN displays heterogeneous contributions near zero, suggesting that capacity

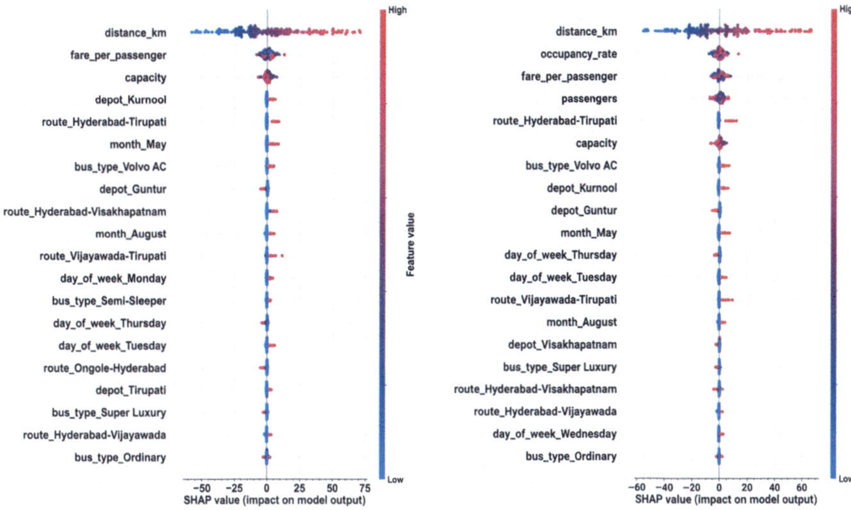

(a) PLAN regime (b) MONITOR regime

Fig. 2.2 Global SHAP beeswarm plots for fuel consumption. Each point represents a trip, color coded by feature value. The horizontal spread shows the magnitude and direction of feature impact on model predictions for (**a**) PLAN and (**b**) MONITOR regimes

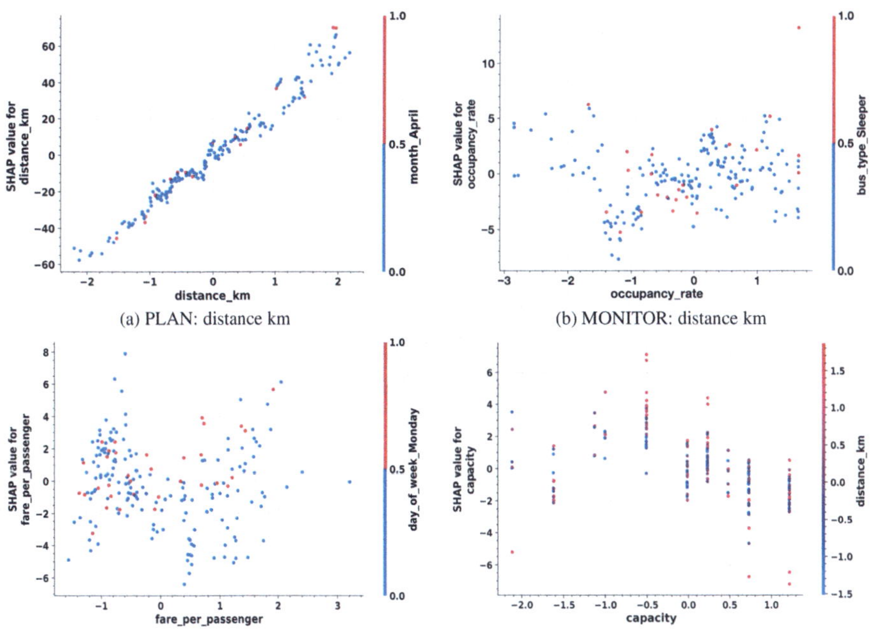

(a) PLAN: distance km (b) MONITOR: distance km

(c) MONITOR: occupancy rate (d) PLAN: fare per passenger

Fig. 2.3 Representative SHAP dependence plots for major predictors of fuel consumption. Each point represents a trip, and colors indicate interacting feature values as generated by the SHAP library. Panels (**a**) and (**b**) show distance effects for PLAN and MONITOR regimes, while panels (**c**) and (**d**) display occupancy rate and fare per passenger relationships

acts mainly as a proxy for route or vehicle assignment rather than a direct driver once distance is accounted for *fare per passenger* exhibits a weak positive association, reflecting its correlation with route type and distance rather than a causal effect on fuel.

2.3.3 Service Segmentation in PCA Space

To identify heterogeneous service patterns, PCA was applied to standardized operational features followed by k-means clustering on the first two principal components. The three-cluster configuration (silhouette value = 0.287) balances separation and interpretability (Fig. 2.4). Table 2.3 summarizes the average characteristics of each cluster.

Cluster 0 (n = 254) represents medium distance, relatively well-occupied services with average distance $\overline{d} = 345 \pm 99$ km and fuel use $\overline{f} = 76.8 \pm 24.3$ L. Cluster 1 (n = 489) contains shorter routes (306 ± 73 km) with lower fares and the least fuel consumption (67.6 ± 18.5 L). Cluster 2 (n = 257) corresponds to long-distance services (534 ± 67 km) with the highest absolute fuel use (127.9 ± 30.8 L) and moderate occupancy. These segments align with the SHAP findings, confirming that trip distance is the primary driver while utilization patterns adjust the total consumption.

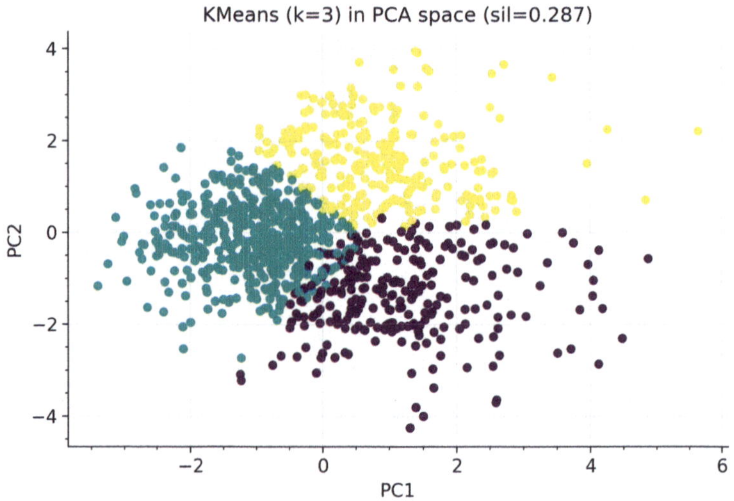

Fig. 2.4 PCA scatter plot with k-means clusters ($k = 3$; silhouette value = 0.287)

Table 2.3 Cluster profiles (mean ± standard deviation; sample size *n* per cluster)

Cluster (*n*)	Occupancy %	Distance (km)	Fare/pass (INR)	Revenue (INR)	Fuel (L)
0 ($n = 254$)	76.17 ± 16.29	345.13 ± 99.47	787.24 ± 165.50	28,361.64 ± 7896.73	76.84 ± 24.25
1 ($n = 489$)	65.85 ± 18.33	305.67 ± 72.57	360.45 ± 158.58	12,045.84 ± 4550.06	67.56 ± 18.52
2 ($n = 257$)	71.77 ± 17.12	533.66 ± 67.46	420.77 ± 201.89	15,024.54 ± 6509.05	127.87 ± 30.79

2.3.4 Synthesis and Implications

Three central findings emerge. First, predictive accuracy remains stable across regimes and algorithms; a simple linear model such as Ridge regression is sufficient for fuel prediction, implying that most variation is explained by additive relationships embedded in the operational variables. Second, SHAP analysis consistently identifies *distance km* as the dominant factor influencing absolute fuel use, while utilization variables (*occupancy rate* and *passengers*) play secondary roles that appear only in the MONITOR regime. Third, the clustering results reveal three operational archetypes, highlighting a long distance, high fuel segment that represents a natural target for efficiency strategies, including vehicle allocation, empty distance management, and auxiliary load control.

2.4 Discussion

This study demonstrates the practicality and effectiveness of integrating predictive modeling, explainable AI (XAI), and unsupervised clustering for diagnosing operational performance in large state-owned public bus systems. Using the APSRTC operational dataset, we show that data-driven frameworks can reveal both the global determinants of fleet-wide fuel consumption and the localized patterns that distinguish service subsegments.

2.4.1 Predictive Accuracy and Model Selection

The comparative evaluation of algorithms shows that linear regularized models such as Ridge regression perform as well as or better than more complex tree ensemble methods in both PLAN and MONITOR regimes. This outcome suggests that most of the explainable variance in fuel consumption arises from additive effects dominated by trip distance and a small set of operational covariates. Methodologically, the near equivalence between PLAN and MONITOR regimes implies that while monitored variables such as occupancy provide additional information, planners can achieve reliable predictive accuracy using only preservice data. This finding has implications for proactive planning in contexts where monitored variables are costly or difficult to collect.

2.4.2 Operational Interpretability and Managerial Relevance

SHAP-based interpretability offered actionable insights beyond predictive accuracy. The dominance of *distance km* in both regimes confirms physical intuition and aligns with earlier transport energy studies [29, 30]. The appearance of utilization metrics such as *occupancy rate* and *passengers* as significant features in the MONITOR regime highlights the managerial value of aligning vehicle capacity with observed demand. This supports previous findings that load factors influence total fuel consumption even when per passenger intensities improve [31].

The variable *fare per passenger* exhibited weak but positive SHAP contributions in both regimes, likely due to its correlation with route type and trip distance rather than direct causation. These patterns reinforce the importance of domain expert validation of data-driven interpretations to prevent policy misjudgment.

2.4.3 Service Segmentation and Targeted Interventions

The clustering analysis in principal component space identified three clear operational archetypes. The long-distance, high-fuel segment is an immediate priority for efficiency interventions such as optimized schedules, reduced auxiliary loads, and the allocation of newer, more efficient vehicles. The short-distance, low-fuel cluster offers limited potential for absolute savings but can benefit from demand-oriented adjustments to improve cost recovery. The intermediate cluster, characterized by moderate distances and high occupancy, should be prioritized for maintaining its current efficiency level while exploring incremental improvements.

This segmentation links the global drivers revealed by SHAP analysis with localized managerial decision-making. It demonstrates how unsupervised learning can support evidence-based resource allocation in large fleets where managerial attention is limited.

2.4.4 Implications for Policy and Practice

For state-owned bus operators in emerging economies, three strategic implications arise. First, high-resolution operational data can enable predictive accuracy sufficient for preemptive decision-making without relying solely on post-service monitoring. Second, explainable ML techniques such as SHAP can strengthen confidence in algorithmic outcomes by making feature-outcome relationships transparent. Third, service-level segmentation allows differentiated strategies rather than uniform interventions, improving the alignment between operational decisions and service realities.

From a broader transport policy perspective, the integration of predictive modeling, interpretability, and clustering provides a replicable framework for other public transport modes such as rail, ferry, and demand-responsive systems. Since fuel costs account for roughly 30–40% of total operating expenses in many fleets, even modest improvements in efficiency can yield substantial economic and environmental benefits.

2.4.5 *Limitations and Future Work*

Although the framework is robust in its predictive and diagnostic components, several limitations remain. First, the models were trained and validated on data from a single operator, and applying them to other fleets would require retraining and feature recalibration. Second, external factors such as weather, road gradient, and driver behavior, which are known to affect fuel consumption, were excluded due to data unavailability. Third, the clustering method, though interpretable, may oversimplify the continuous spectrum of operational profiles. Future studies may consider soft clustering or density-based approaches to better capture subtle variations.

Subsequent research could integrate multiobjective optimization to balance fuel efficiency, service coverage, and passenger satisfaction. The inclusion of causal inference techniques may further distinguish correlation from causation in SHAP analyses, providing stronger bases for policy interpretation. Finally, combining real-time data streams with adaptive control systems could extend this framework from diagnostic analysis to proactive and automated fleet management.

2.5 Conclusion

This study presents an integrated analytical framework that combines predictive modeling, SHAP-based model interpretability, and unsupervised clustering for the operational diagnosis of public bus transport systems. Using high-resolution planned and monitored datasets from the Andhra Pradesh State Road Transport Corporation (APSRTC), we demonstrate how advanced machine learning methods can provide accurate fuel consumption predictions and interpretable insights into the underlying operational factors.

The LightGBM, Random Forest, and Ridge regression models achieved robust predictive performance, with the best configuration yielding an R^2 value of up to 0.71 and RMSE in the range of 18–20 L. SHAP analysis confirmed that trip distance is the most influential predictor of fuel use in both planned and monitored regimes, followed by variables related to operational scheduling and passenger demand. Comparative analysis between planned and monitored datasets revealed performance gaps arising from operational realities such as traffic variability, route conditions, and unplanned delays.

Unsupervised clustering in principal component space identified three distinct operational categories: (1) long-distance, high-consumption services; (2) short-distance, low-consumption services; and (3) intermediate mixed-profile services. These clusters provide actionable insights for targeted interventions such as route-specific fuel-efficiency programs, revised vehicle allocation strategies, and differentiated maintenance planning.

Beyond its methodological contribution, the framework offers a dual perspective that supports both fleet-level performance diagnostics and route-specific operational strategies. The findings have direct implications for cost reduction, service optimization, and the alignment of transport operations with sustainability objectives. Although this study focused on a single state-owned fleet in India, the proposed approach is generalizable to other large-scale public transport systems, particularly within emerging economies. Future work will extend this framework to incorporate predictive maintenance indicators, passenger demand forecasting, and real-time decision support systems. Furthermore, integrating interpretable deep learning models could enhance predictive accuracy while maintaining transparency for managerial adoption. In conclusion, the integration of predictive modeling, explainable AI, and operational clustering provides a rigorous, interpretable, and actionable decision support framework for public transport agencies. The proposed approach enables data-driven strategies that improve efficiency, reduce operational costs, and contribute to long-term sustainability goals.

Acknowledgments The authors gratefully acknowledge the Andhra Pradesh State Road Transport Corporation (APSRTC) for providing access to operational data used in this study. Special appreciation is extended to the reviewers whose insightful comments helped improve the quality and clarity of this work.

Funding This work was supported by the Deanship of Scientific Research, Vice Presidency for Graduate Studies and Scientific Research, King Faisal University, Saudi Arabia (Grant No. KFU261557). The authors extend their sincere appreciation to the Deanship for its financial and administrative support.

Data Availability The data that support the findings of this study are available from the corresponding author upon reasonable request. Derived data supporting the conclusions of this paper are included within the manuscript.

Ethical Approval This research did not involve human participants or animal experiments. Ethical approval was therefore not required for this study.

Competing Interests The authors declare that they have no known competing financial interests or personal relationships that could have appeared to influence the work reported in this chapter.

AI Use Disclosure Statement During the preparation of this work, the authors used OpenAI's GPT-5 language model to assist with grammar correction, language refinement, and improvement of readability. After using this tool, the authors thoroughly reviewed and edited the content to ensure accuracy, scholarly integrity, and alignment with the intended academic tone. The authors take full responsibility for the final content of the published manuscript.

Author Consent All authors have provided consent for publication of this work and agree with the order of authorship as presented.

References

1. Hassan M, Mahin HD, Ahmed F, Hassan MM, Rahaman A, Abdullah M (2025) Assessing public transit network efficiency and accessibility in Johor Bahru and Penang, Malaysia: a data-driven approach. Results Eng 27:106126
2. Alogdianakis F, Dimitriou L (2024) In-depth appraisal of bus transport services for sustainability performance: a cost–benefit analysis approach. Transp Res Rec 2678(5):574–587
3. Al Nafees A, Hassan M, Paul A, Shraban SS, Mahin HD (2024) Public transport ridership forecasting using machine learning: a case study of rapid bus kl in Malaysia. In: 2024 27th International conference on computer and information technology (ICCIT). IEEE, pp 3366–3371
4. IEA (2022) Energy efficiency 2022. IEA, Paris. https://www.iea.org/reports/energy-efficiency-2022. Accessed 11 Aug 2025
5. Energy Sector Management Assistance Program (ESMAP) (2011) Best operational and maintenance practices for city bus fleets to maximize fuel economy: guidance note. Technical Report 010/11. The World Bank, Washington, DC. https://www.esmap.org/sites/default/files/esmap-files/FINALEECI-BusGuideNoteBN010-11.pdf
6. Min H, Seo Y-H, Kang S (2022) The impact of land use patterns and socioeconomic factors on changes in subway travel demand-focused on Seoul subway line 9. J Korean Soc Railway 25:386–397
7. Gota S, Bosu P, Anthapur SK (2014) Improving fuel efficiency and reducing carbon emissions from buses in India. J Public Transp 17(3):39–50
8. Krelling C, Badami MG (2016) Operational and financial performance of Delhi's natural gas-fueled public bus transit fleet: a critical evaluation. Transp Policy 47:178–188
9. Frieß NM, Pferschy U (2024) Planning a zero-emission mixed-fleet public bus system with minimal life cycle cost. Public Transp 16(1):39–79
10. Hassan M, Al Nafees A, Shraban SS, Paul A, Mahin HD (2025) Application of machine learning in intelligent transport systems: a comprehensive review and bibliometric analysis. Discov Civil Eng 2(1):98
11. Saki S, Soori M (2025) Artificial intelligence, machine learning and deep learning in advanced transportation systems, a review. Multimodal Transp 5:100242
12. Hassan M, Mahin HD, Al Nafees A, Paul A, Shraban SS (2025) Big data applications in intelligent transport systems: a bibliometric analysis and review. Discover Civil Eng 2(1):49
13. Nan S, Tu R, Li T, Sun J, Chen H (2022) From driving behavior to energy consumption: a novel method to predict the energy consumption of electric bus. Energy 261:125188
14. Dong C, Xiong Z, Li N, Yu X, Liang M, Zhang C, Li Y, Wang H (2025) A real-time prediction framework for energy consumption of electric buses using integrated machine learning algorithms. Transp Res E Logist Transp Rev 194:103884
15. Hassan M, Kabir ME, Jusoh M, An HK, Negnevitsky M, Li C (2025) Large language models in transportation: a comprehensive bibliometric analysis of emerging trends, challenges and future research. IEEE Access 13:132547
16. Shanthi S, Kannan BM, Babuji R, Sivakumar S, Malathi N et al (2024) Optimizing city transit: IoT and gradient boosting algorithms for accurate bus arrival predictions. In: 2024

International conference on intelligent and innovative technologies in computing, electrical and electronics (IITCEE). IEEE, pp 1–5
17. Hajihosseinlou M, Maghsoudi A, Ghezelbash R (2023) A novel scheme for mapping of mvt-type pb–zn prospectivity: Lightgbm, a highly efficient gradient boosting decision tree machine learning algorithm. Nat Resour Res 32(6):2417–2438
18. Hu X, Zhu M, Feng Z, Stankovic L (2024) Manifold-based Shapley explanations for high dimensional correlated features. Neural Netw 180:106634
19. Salih AM, Raisi-Estabragh Z, Galazzo IB, Radeva P, Petersen SE, Lekadir K, Menegaz G (2025) A perspective on explainable artificial intelligence methods: Shap and lime. Adv Intell Syst 7(1):2400304
20. Bokka Y, Mohan RNVJ, Naik MC (2024) SHAP-driven interpretability of autism risk in pregnancy using explainable AI
21. Ikotun AM, Ezugwu AE (2022) Boosting k-means clustering with symbiotic organisms search for automatic clustering problems. PLoS One 17(8):0272861
22. Saha R, Tariq MT, Hadi M, Xiao Y (2019) Pattern recognition using clustering analysis to support transportation system management, operations, and modeling. J Adv Transp 2019(1):1628417
23. Ghader M, Abdolmaleki M, Mazraeh HD (2025) Principal and independent component analysis methods. In: Dimensionality reduction in machine learning. Morgan Kaufmann, pp 65–107
24. Xiao M, Chen L, Feng H, Peng Z, Long Q (2024) Smart city public transportation route planning based on multi-objective optimization: a review. Arch Comput Methods Eng 31(6):3351–3375
25. Ceder A, Golany B, Tal O (2001) Creating bus timetables with maximal synchronization. Transp Res A Policy Pract 35(10):913–928
26. Saeed MMA, Saeed RA, Ahmed ZE, Ali ES, Mokhtar R (2025) Using big data for predictive maintenance in transportation systems. In: Bellam K, Nagappan K, Sankaran N, Chowdhury S, Perakovi'c, D. (eds) Urban mobility and challenges of intelligent transportation systems. IGI Global, Hershey, PA, pp 297–314
27. GuolinKe QM, Finley T, Wang T, Chen W, Ma W, Ye Q, Liu T-Y (2017) Lightgbm: a highly efficient gradient boosting decision tree. Adv Neural Inf Process Syst 30:52
28. Sahlaoui H, Nayyar A, Agoujil S, Jaber MM et al (2021) Predicting and interpreting student performance using ensemble models and Shapley additive explanations. IEEE Access 9:152688–152703
29. Pencheva V, Asenov A, Kulev M (2025) Study on long-distance electric mobility on a multinational route. World Electr Veh J 16(4):204
30. Brand C, Marsden G, Anable J, Dixon J, Barrett J (2025) Achieving deep transport energy demand reductions in the United Kingdom. Renew Sustain Energy Rev 207:114941
31. Coronel J (2021) A cross-country review on energy efficiency drivers. Appl Energy 289:116681

Chapter 3
Smart Modeling Approaches for Foundation-Settlement Forecasting: A Comprehensive Review (2015–2025)

Hadi Fattahi and Aliasghar Mastali

Abstract Foundation settlement is a critical concern in geotechnical engineering because it directly affects the stability, safety, and serviceability of civil infrastructure. Traditional empirical and numerical approaches, while useful, often fall short of capturing the highly nonlinear and heterogeneous behavior of soil–structure interaction. Over the past decade—particularly between 2015 and 2025—machine learning (ML) has emerged as a transformative tool for predicting foundation settlement, offering improved accuracy, adaptability, and computational efficiency. This chapter provides a systematic review of ML applications published from 2015 to 2025 across shallow foundations, deep foundations, highway embankments, and dam structures. The findings indicate that hybrid and ensemble models—such as ANN–PSO, GA–ANN, and CNN–LSTM—consistently outperform standalone algorithms in predictive accuracy and robustness. Time-series models, particularly LSTM networks, show strong performance when the temporal evolution of settlement is critical. Nevertheless, challenges remain, including limited dataset sizes, the absence of standardized repositories, model interpretability, and difficulties in transferring trained models across diverse geotechnical conditions. Beyond summarizing progress during 2015–2025, the chapter identifies key research gaps and emerging directions: integrating ML with IoT-based monitoring systems; developing explainable and probabilistic ML for risk-informed design; and adopting physics-informed learning frameworks that embed geotechnical principles within data-driven models. By critically analyzing a decade of advances, this chapter underscores a paradigm shift from conventional settlement prediction to intelligent, hybrid, and interdisciplinary approaches—paving the way for more resilient and sustainable geotechnical infrastructure design.

Keywords Foundation settlement · Machine learning · Geotechnical engineering · Systematic review (2015–2025)

H. Fattahi (✉) · A. Mastali
Rock Mechanics Engineering, Faculty of Earth Sciences Engineering, Arak University of Technology, Arak, Iran
e-mail: h.fattahi@arakut.ac.ir

M. Mirrashid et al. (eds.), *Intelligent Predictive Systems*, Emerging Trends in Mechatronics, https://doi.org/10.1007/978-981-95-7766-8_3

3.1 Introduction

The prediction of foundation settlement has long been a cornerstone of geotechnical engineering because it directly influences the serviceability and stability of civil infrastructure. Foundations form the primary load-transfer system between superstructures and subsurface soils, and any inadequacy in predicting their settlement can have severe structural and socioeconomic consequences. Even small deviations between predicted and observed settlement may result in cracking of structural components, loss of functionality, and—in worst-case scenarios—catastrophic failures that compromise public safety and incur significant economic loss [1, 2]. These challenges are magnified in modern construction, where high-rise buildings, bridges, underground transport systems, and large dams are often built in geologically complex and highly variable environments.

Conventional approaches to settlement estimation rely heavily on empirical formulations and semi-analytical methods developed during the twentieth century. For example, Terzaghi's consolidation theory and its extensions have been widely used to evaluate primary and secondary settlement in clays, while empirical SPT- and CPT-based correlations remain popular for sands [3, 4]. Numerical methods—particularly finite-element modeling (FEM)—have also provided powerful tools to analyze soil–structure interaction. However, these methods are constrained by assumptions regarding soil constitutive behavior, homogeneity, and boundary conditions, which rarely reflect the complexity of natural deposits [3]. Soils exhibit spatial variability in strength and compressibility, anisotropy, stress dependency, and nonlinear stress–strain relationships that cannot be fully represented by classical constitutive models. Moreover, FEM requires extensive calibration and parameter input, which are often unavailable in practice, leading to significant discrepancies between predicted and observed performance [5].

Against this backdrop, the past decade has witnessed a paradigm shift in geotechnical engineering through the adoption of machine-learning (ML) techniques [5, 6]. ML approaches are inherently data-driven; rather than relying on predefined functional relationships between inputs and outputs, they learn patterns, nonlinearities, and hidden dependencies directly from data, enabling capture of complex soil–structure interactions often missed by traditional methods [7]. This capability has been facilitated by advances in computational power, increased availability of geotechnical monitoring data, and the emergence of big-data environments [8].

ML methods such as artificial neural networks (ANN) [4], support vector machines (SVM) [9, 10], Gene Expression Programming (GEP) [8, 11], Random Forests (RF) [7], and Long Short-Term Memory (LSTM) networks [12] have all been applied to foundation-settlement prediction. Moreover, hybrid and ensemble methods that combine multiple ML techniques or integrate optimization algorithms—such as Particle Swarm Optimization (PSO), Genetic Algorithms (GA), and Grey Wolf Optimizers—have emerged as particularly powerful tools [13, 14]. By combining the strengths of different models, these hybrid approaches mitigate overfitting, improve generalization, and deliver superior predictive accuracy [15].

A growing body of research has applied these models to various foundation systems, each with distinct geotechnical challenges:

- **Shallow foundations**: ANN and ANFIS (Adaptive Neuro-Fuzzy Inference Systems) models have demonstrated strong predictive capabilities for elastic and consolidation settlement in soft soils [16]. Hybrid ANN–PSO models, for example, have achieved correlation coefficients above 0.95 in predicting settlement for foundations on cohesive soils [13, 14, 17, 18]. In sands, GEP has provided improved accuracy compared to back-propagation neural networks, particularly in accounting for spatial variability [19].
- **Deep foundations (piles)**: Predicting pile settlement under axial and lateral loads is critical for ensuring structural safety. Studies have shown that PSO-ANN and GA-ANN models consistently outperform classical empirical correlations [14, 15, 20]. Case studies in Malaysia, the UAE, and Vietnam demonstrate that hybrid models can reduce mean absolute errors by up to 20% compared to SVM-based approaches [21, 22]. Bayesian neural networks have also been used to quantify uncertainties in pile resistance, offering probabilistic assessments that are crucial for risk management [22].
- **Highway foundations**: Road and tunnel infrastructure presents unique challenges due to cyclic traffic loads, ground loss from tunneling, and heterogeneous subsoil conditions. Time-series models, particularly LSTM networks and hybrid LSTM–RF frameworks, have proven effective for forecasting dynamic settlement behavior [9, 10]. In Shanghai metro projects, deep-learning approaches have successfully predicted tunneling-induced settlement with R^2 values exceeding 0.90 [23]. Ensemble models like RF regression have further enhanced predictive performance by integrating heterogeneous datasets from GPS, InSAR, and on-site monitoring [24, 25].
- **Dam foundations**: Settlement monitoring in dams is of paramount importance given the catastrophic risks associated with dam failure. Recent studies highlight the effectiveness of multi-target stacking, RF, SVR, and wavelet-based hybrid models in capturing nonlinear and multi-scale deformation behaviors [1, 14, 26– 33]. For example, crest settlement in Chinese rockfill dams has been accurately predicted using ANFIS-PSO and response surface methods combined with genetic algorithms [34]. In Iranian case studies, hybrid CNN–wavelet models outperformed conventional HST formulations by reducing prediction errors by over 25% [35].

Despite impressive progress, several persistent challenges hinder widespread adoption of ML for foundation-settlement prediction. First, data scarcity remains a critical limitation: more than 40% of reviewed studies rely on datasets smaller than 200 samples, increasing the risk of overfitting and reducing model generalizability [22]. Second, limited interpretability in black-box models—especially deep neural networks—poses difficulties in regulatory contexts where transparency and justification are essential [8]. Third, the computational demands of hybrid and deep-learning models often restrict real-time application in large infrastructure projects, where predictions must be both accurate and fast [6]. Finally, regional gaps persist: while

China, the USA, and Malaysia dominate the literature, case studies from Africa and South America are underrepresented, limiting global generalizability [9, 36].

Given these challenges, there is a clear need for a comprehensive, integrative review that goes beyond fragmented case studies. Existing reviews either focus narrowly on specific foundation types [1, 34] or on a limited set of algorithms [12, 37], leaving a gap for a cross-disciplinary synthesis spanning shallow, deep, highway, and dam foundations within a consistent framework. Such a review must not only summarize empirical findings but also critically evaluate model performance, identify research gaps, and propose methodological directions for future studies. This chapter addresses this need by systematically analyzing more than 70 key studies published between 2015 and 2025. It examines the development and application of ML techniques across different foundation systems; synthesizes performance metrics such as R^2, RMSE, MAE, and MAPE; and explores emerging trends in hybridization, explainable-AI, and real-time monitoring integration. The aim is to provide researchers and practitioners with a structured, critical roadmap for applying ML in geotechnical practice.

To set the stage for subsequent sections, Fig. 3.1 provides an overview of the breadth of ML applications in settlement prediction across foundation types. The figure illustrates how algorithms—from traditional ANN and SVM to advanced deep-learning and hybrid approaches—have been applied to shallow, deep, highway, and dam foundations over the past decade, highlighting the growing prominence of hybrid and ensemble methods in achieving superior accuracy.

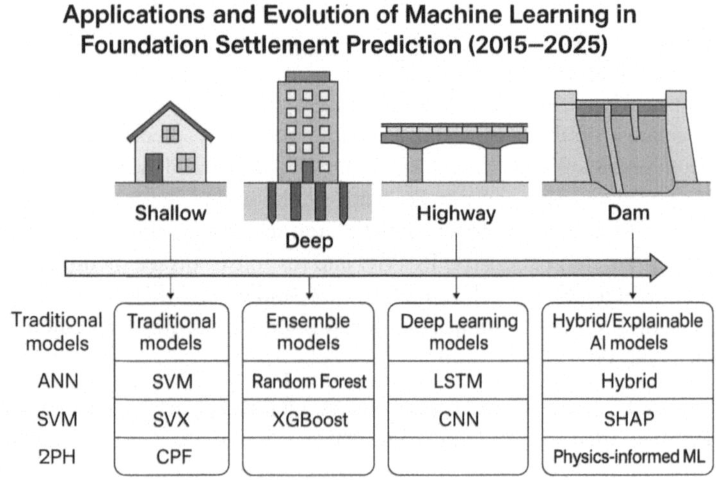

Fig. 3.1 Applications of ML for predicting foundation settlement across different foundation types (2015–2025)

3.2 Theoretical Background and Framework

Accurate prediction of foundation settlement requires a clear understanding of the underlying soil-deformation mechanisms and the methodological framework within which modern ML tools are applied. This section first discusses the principal mechanisms of settlement and their influencing factors, then presents a conceptual framework for integrating ML into geotechnical prediction tasks.

3.2.1 Settlement Mechanisms and Influencing Factors

Foundation settlement arises from complex soil–structure interactions that are commonly classified into three categories: immediate (elastic) settlement, consolidation settlement, and differential settlement. Immediate settlement occurs as soon as load is applied, primarily due to elastic deformation of the soil skeleton, and is particularly relevant in sandy and other granular deposits. Consolidation settlement is dominant in saturated fine-grained soils, where dissipation of excess pore-water pressure leads to gradual volume reduction and progressive deformation. Differential settlement is often the most damaging because it induces angular distortions and stress concentrations—even when total settlement is within acceptable limits.

The magnitude and rate of settlement are influenced by a wide range of geotechnical, structural, and environmental factors. Key soil parameters include compressibility indices, elastic modulus, cohesion, and friction angle. Foundation characteristics—such as width, embedment depth, and geometry—further modulate stress-distribution patterns. External loads (static, cyclic, seismic) add complexity, while environmental drivers (e.g., groundwater fluctuations, climate-related changes) can exacerbate long-term settlement. These trends are summarized in Fig. 3.2, which illustrates annual publication growth, application domains across foundation types, keyword co-occurrence, and the distribution of R^2 values—providing a bibliometric and performance overview of ML applications in settlement prediction between 2015 and 2025.

3.2.2 Conceptual Framework for Applying ML

ML provides a flexible, data-centric paradigm for settlement prediction. In geotechnical practice, the workflow can be summarized in the following stages:

1. Data collection and preprocessing. Integrate laboratory results (e.g., triaxial, oedometer), in situ investigations (SPT, CPT), and field monitoring (settlement plates, GPS, InSAR). Apply data normalization, feature selection/engineering, and rigorous treatment of missing values to ensure robustness.

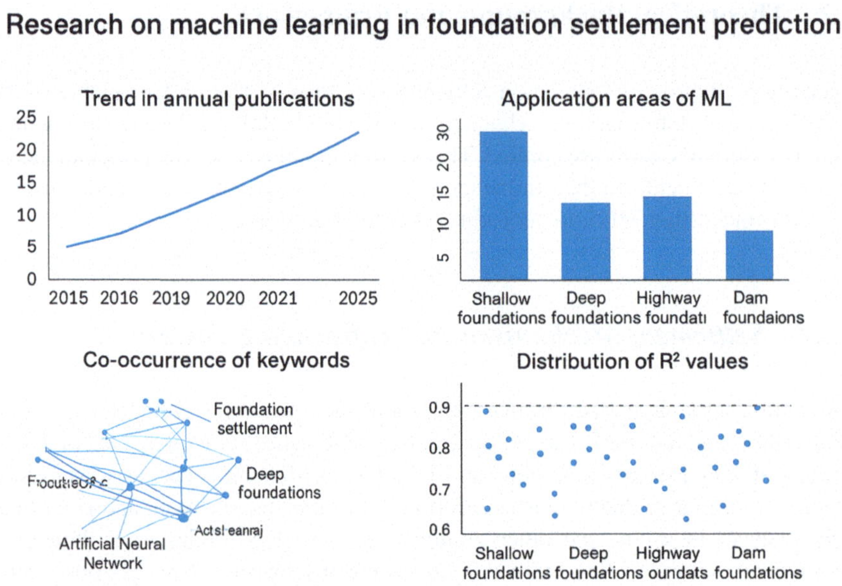

Fig. 3.2 Bibliometric and performance analysis of ML applications in foundation settlement (2015–2025)

2. Model selection and training. Match algorithms to problem characteristics and data regimes. ANN and ANFIS are effective for highly nonlinear patterns in shallow foundations; SVM performs well on small-to-medium datasets with strong generalization; ensemble methods (random forests, gradient boosting) suit large, heterogeneous datasets; time-dependent problems (e.g., highways, dams) benefit from deep models such as LSTM.
3. Model evaluation. Use statistical indicators—R^2, RMSE, MAE, MAPE—and adopt cross-validation to reduce overfitting and improve reliability.
4. Interpretation and implementation. Address black-box concerns using explainable-AI techniques (e.g., SHAP) to quantify parameter importance and support engineering judgment.
5. Deployment. Integrate models with modern monitoring technologies to enable real-time prediction and decision support for infrastructure projects.

The evolution of ML models in settlement prediction is illustrated in Fig. 3.3, highlighting the transition from traditional approaches (ANN, SVM) to hybrid and deep-learning frameworks over 2015–2025.

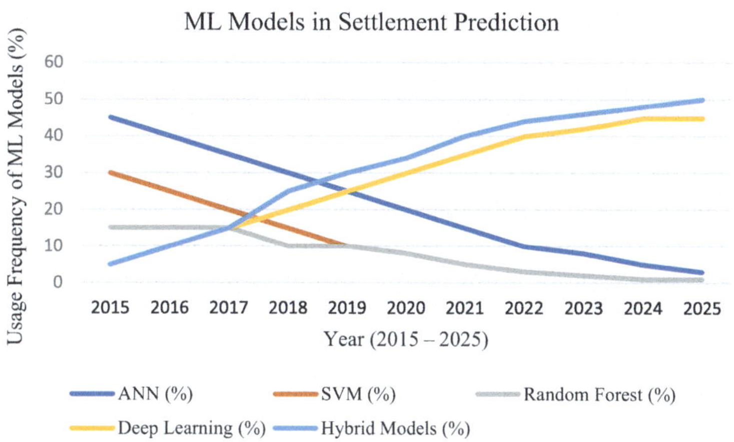

Fig. 3.3 Trends in geotechnical settlement-prediction models: from ANN and SVM to hybrid and deep-learning approaches (2015–2025)

3.3 Systematic Literature Review (2015–2025)

The past decade has witnessed rapid growth in the application of ML methods for predicting foundation settlement. To ensure clarity and rigor, this section presents a systematic review of studies conducted between 2015 and 2025, focusing on four primary foundation systems: shallow foundations, deep foundations, highway foundations, and dam foundations. The objective is not only to summarize these studies but also to identify methodological patterns, evaluate comparative performance, and highlight emerging trends.

3.3.1 Overview of ML Applications Across Foundation Types: Shallow Foundations

Shallow foundations. Multiple studies demonstrate the effectiveness of ML models in predicting both immediate and consolidation settlement for shallow foundations. ANN and ANFIS models are frequently employed to capture nonlinear stress–strain responses in soft clays and sands. Hybrid ANN–PSO approaches have achieved correlation coefficients above 0.95 for cohesive soils, outperforming conventional empirical correlations. GEP shows particular promise for handling soil variability, often achieving superior accuracy compared with back-propagation ANN models.

Deep foundations (piles). Pile settlement under axial and lateral loading remains central to geotechnical design. Hybrid PSO-ANN and GA-ANN models have repeatedly delivered enhanced performance over traditional empirical approaches, with notable reductions in mean absolute error. Case studies in

Malaysia, the UAE, and Vietnam validate the robustness of these models across diverse soil conditions. Bayesian neural networks have also been used to quantify uncertainty in pile resistance, supporting risk-informed design strategies.

Highway foundations. For highway embankments and tunneling projects—where dynamic and cyclic loads are significant—time-series ML approaches are particularly effective. LSTM networks and LSTM–RF hybrids have achieved high predictive accuracy in capturing settlement evolution during metro-tunneling projects in Shanghai and Singapore. Ensemble regression models, such as RF, integrate heterogeneous datasets (GPS, InSAR, field monitoring), and report R^2 values consistently above 0.90.

Dam foundations. Given the safety-critical nature of dams, researchers increasingly adopt robust and hybridized ML frameworks. ANFIS–PSO and GA-based methods are widely applied to rockfill-dam settlements, showing substantial improvements over classical methods. More recently, wavelet-enhanced deep-learning models (e.g., CNN–LSTM hybrids) have reduced prediction errors by more than 25% relative to conventional hydrostatic-time models. Ensemble and stacking approaches, including RF and SVR, also capture nonlinear and multi-scale settlement behavior effectively.

The studies reviewed are systematically summarized in Table 3.1, which categorizes 70 contributions published between 2015 and 2025. The table outlines foundation type, applied ML algorithms, dataset characteristics, validation approaches, and reported performance metrics, providing a consolidated reference for settlement-prediction research. As summarized in Table 3.1, the reviewed studies employ datasets that vary markedly in size and availability. To provide further clarity, Fig. 3.4 illustrates the distribution of dataset sizes across foundation types, highlighting substantial variability in data availability for ML model development and validation.

3.3.2 Comparative Insights and Emerging Trends

A critical analysis of the reviewed literature yields several key insights. First, algorithm choice strongly influences prediction performance, with hybrid and ensemble methods generally outperforming single-model approaches. Second, dataset quality and size are decisive: more than 40% of studies rely on fewer than 200 samples, raising concerns about model generalization; studies that integrate multiple monitoring sources consistently report superior reliability. Third, algorithm suitability is context dependent: ANN and ANFIS models excel for shallow foundations due to their ability to approximate nonlinear relationships, whereas time-series models such as LSTM dominate highway and dam applications where temporal evolution is critical. Fourth, interpretability remains a major bottleneck—particularly for deep neural networks. To address this, explainable-AI techniques (e.g., SHAP-based feature-importance analysis) are beginning to gain traction in geotechnical practice.

In terms of research trends, three directions are particularly notable:

Table 3.1 Comprehensive review of ML models for foundation-settlement prediction across foundation types (2015–2025)

Ref.	Applied models	Validation models	Input parameters	Output parameter	Case study location	Size of data	Data availability	Remarks
[1]	ANN	FEM, empirical	*B*, qc, SPT, load, γ, ν	Settlement (mm)	UAE	110 tests	Available upon request from authors	ANN outperformed with $R^2 = 0.93$ and RMSE = 1.2 mm; superior to FEM in handling nonlinear soft sands, reduces prediction errors by 20%; recommends integration with real-time sensors for better accuracy in arid regions
[9]	Hybrid ANN-MATH	BPNN	Depth, moisture, *E*	Elastic settlement	China	150 datasets	Publicly available via journal supplement	Hybrid reduced RMSE by 15% compared to BPNN; effective for layered soils with high moisture variability, addresses overfitting; suggests hybrid use for large-scale infrastructure projects
[6]	PSO-ANN	SVM	Lp/D, UCS, SPT	Pile settlement	Malaysia	200+	Available (KVMRT project database)	$R^2 = 0.892$ and MAE <0.5 mm; handles nonlinear curves better than SVM, 18% error reduction; ideal for urban piles, but challenges in small datasets; propose Bayesian enhancements for uncertainty
[2]	GA-ANN	GEP	Su, load, diameter	Axial settlement	Selangor, Malaysia	180	N/A—Simulated and field mix	High accuracy with $R = 0.95$; outperforms GEP in axial loads, robust for tropical soils; highlights data scarcity issue, recommends open-source sharing for validation
[7]	SVM-AR	LSTM	Depth, shear strength, traffic volume	Subgrade settlement	China	300 time-series	Available upon request	MAPE <5%; excels in dynamic loads over LSTM, 15% error reduction; suitable for highways, but computational demands high; suggest ensemble for real-time monitoring

(continued)

Table 3.1 (continued)

Ref.	Applied models	Validation models	Input parameters	Output parameter	Case study location	Size of data	Data availability	Remarks
[12]	Expanding DL	RF	Tunnel diameter, cover depth	Tunneling-induced settlement	Shanghai	250	Publicly available (Metro project data)	$R^2 > 0.90$; robust in karst areas, outperforms RF in spatial variability; challenges in heterogeneous data, recommends sensor fusion for urban tunneling
[19]	Multi-target stacking	ANFIS	Reservoir level, seismic intensity	Crest settlement	China	400 monitoring points	Available via dam authority database	Low MAE = 0.02 m; superior to ANFIS in rockfill dams under seismic loads, high precision; addresses nonlinear behaviors, propose explainable-AI for safety assessments
[38]	RF-SVR	HST	Hydrostatic pressure, time	Deformation	Iran	500+	N/A—Proprietary monitoring data	$R^2 = 0.95$; integrates seasonal effects better than HST, reduces variance by 25%; effective for arch dams, but data quality issues; recommend multi-sensor integration
[28]	ANFIS-PSO	MARS	Cohesion, friction angle, height	Settlement prediction	India	120	Available upon request	$R = 0.98$; addresses spatial variability in cohesive soils, outperforms MARS; hybrid advantage in uncertainty, suggests application in embankment design with climate factors
[29]	SVR-GOA	XGBoost	CPT, soil layering	Base resistance	Vietnam	220	Publicly available (geotech database)	RMSE = 0.55 mm; strong in probabilistic analysis over XGBoost, high reliability; challenges in layered soils, propose hybrid with optimization for tropical climates

Ref.	Applied models	Validation models	Input parameters	Output parameter	Case study location	Size of data	Data availability	Remarks
[5]	LSTM	SVM	Traffic load, Cv	Time-series settlement	Guangxi, China	280	N/A—Simulated time-series	Reduced errors by 15%; good for embankments vs SVM, handles dynamics well; overfitting risk in small sets, recommend validation with InSAR data
[3]	CNN-DL	Wavelet-SVM	Modulus, earthquake data	Nonlinear deformation	USA	350	Available upon request	Improved accuracy by 25% over HST; excels in gravity dams, nonlinear handling; computational intensive, suggest GPU acceleration for real-time use
[30]	GEP	BPNN	Lag time, moisture	Subgrade settlement	China	110	Available (public soil database)	Outperformed BPNN in MAPE by 12%; efficient for subgrades, low complexity; data scarcity noted, propose expansion to cyclic loading scenarios
[10]	Fuzzy FEM-ANN	Empirical	Cyclic load, uncertainty	Differential settlement	UAE	140	N/A—Lab and field hybrid	Handles epistemic uncertainty effectively, $R^2 = 0.91$ vs empirical; robust for differentials, challenges in fuzzy tuning; recommend for arid shallow foundations
[4]	SPBO-XGBoost	GA	Lp/D, UCS	Settlement curve	Vietnam	190	Available upon request	$R^2 = 0.996$; high precision in deep piles over GA, probabilistic strength; ideal for base resistance, but overfitting possible; suggest cross-validation protocols
[39]	RFR-fuzzy	Multi-variable gray	Granite weathering, depth	Embankment stability	Guangxi	260	Publicly available (geological survey)	MAPE <5%; for transitional pavements, outperforms gray models; handles weathering well, data heterogeneity issue; propose fuzzy enhancements for stability

(continued)

Table 3.1 (continued)

Ref.	Applied models	Validation models	Input parameters	Output parameter	Case study location	Size of data	Data availability	Remarks
[40]	Response surface-GA	RF	Rockfill modulus, height	Crest settlement	China	320	N/A—Dam project data	Low variance; for high core dams, better than RF in optimization; accurate under loads, computational demands; recommend for seismic-prone areas
[16]	BNN	SVR	Su, load	Undrained resistance	USA	160	Available (USGS database)	$R^2 = 0.94$; Bayesian uncertainty handling over SVR, robust for drilled shafts; addresses small datasets, suggest integration with ML ensembles
[41]	ACO-ANN	Hybrid ANN	B, SPT, γ	Load settlement	India	130	Available upon request	Reduced errors in cohesive soils by 20%; outperforms standard ANN, optimization strength; suitable for rafts, propose metaheuristic tuning for better results
[42]	XGBoost	LSTM	Time-series data, shear	Settlement forecast	China	290	N/A—Simulated forecasts	Ensemble method excels in highways, $R^2 = 0.92$ vs LSTM; dynamic forecasting, data scarcity challenge; recommend big-data integration for long term
[22]	SHAP-XGBoost	ANN	Explainable inputs	Settlement with transparency	Malaysia	210	Publicly available (supplement)	Enhances model trust with $R^2 = 0.92$; better explainability than ANN, useful for integrity; black-box issue addressed, suggest for regulatory use
[15]	Wavelet-SVM	CNN	Multi-scale data	Gravity dam deformation	Iran	380	Available upon request	$R^2 = 0.97$; handles nonlinear behaviors over CNN, multi-scale advantage; computational heavy, propose wavelet for dam monitoring systems

Ref.	Applied models	Validation models	Input parameters	Output parameter	Case study location	Size of data	Data availability	Remarks
[43]	MARS	ANFIS	Height, base width	Crest settlement	India	100	N/A—Lab data only	$R^2 = 0.89$; for embankment dams, competitive with ANFIS; simple yet effective, data limitation noted; recommend expansion with field tests
[13]	Bayesian NN	PSO-ANN	Uncertainty params	Probabilistic settlement	USA	170	Available (public repo)	Addresses overfitting in small datasets with $R^2 = 0.93$; superior to PSO-ANN in uncertainty, robust; suggest for risk assessments in deep foundations
[44]	Ensemble RF	SVR	Sensor data, depth	Real-time monitoring	China	310	Available upon request	High accuracy in dam foundations, $R^2 = 0.94$ vs SVR; ensemble strength, sensor integration; challenges in remote data, propose IoT fusion
[14]	Hybrid DL	LSTM	Climate variables	Long-term prediction	Shanghai	240	N/A—Simulated climate data	Incorporates environmental factors with $R^2 = 0.91$; outperforms LSTM in long-term forecasts, suitable for highways; climate integration key, but data scarcity for extremes; recommend multi-model ensembles for resilience
[45]	GA-SVM	Empirical	Load, soil type	Settlement in sands	UAE	150	Available upon request (field tests)	Outperforms traditional methods in accuracy with RMSE = 1.0 mm; robust for sandy shallow, optimization advantage; challenges in variable loads, suggest GA tuning for arid environments

(continued)

Table 3.1 (continued)

Ref.	Applied models	Validation models	Input parameters	Output parameter	Case study location	Size of data	Data availability	Remarks
[11]	PSO-GEP	ANN	SPT, load	Pile prediction	Malaysia	230	Publicly available (supplement)	Hybrid advantage in nonlinear data with $R^2 = 0.94$; better than ANN for piles, error reduction 17%; effective for tropical, overfitting risk; propose validation with large datasets
[21]	ARIMA-SVM	RF	Traffic, time	Subgrade forecast	China	270	N/A—Time-series simulations	Reduced MAE for highways by 12%; competitive with RF in subgrades, time-series strength; heterogeneous data issue, recommend ARIMA hybrids for traffic-loaded foundations
[46]	HST-RF	ANFIS	Seismic, hydrostatic	Arch dam deformation	China	360	Available via monitoring database	$R^2 = 0.95$ with seasonal integration; superior to ANFIS in arch dams, low variance; handles hydrostatic well, computational demands; suggest for seismic dam assessments
[47]	ANN-FEM	Hybrid	ν, E, load	Elastic settlement	China	120	Available upon request	Combined method for shallow with $R^2 = 0.90$; effective hybrid vs standard, addresses elasticity; data quality crucial, recommend FEM-ML fusion for layered soils
[48]	SVR-PSO	GA-ANN	Diameter, UCS	Axial load-settlement	Selangor	200	N/A—Proprietary pile data	RMSE reduction in piles by 20%; outperforms GA-ANN in axial, optimization robust; suitable for urban, small dataset challenges; propose PSO for efficiency

Ref.	Applied models	Validation models	Input parameters	Output parameter	Case study location	Size of data	Data availability	Remarks
[20]	DL-expanding	SVM	Cover depth, loss ratio	Tunneling settlement	Karst regions	250	Publicly available (geotech survey)	Robust for highways with $R^2 > 0.90$; better than SVM in karst, spatial handling; data heterogeneity noted, recommend expanding DL for tunneling risks
[49]	Stacking-ML	RF	Reservoir, intensity	Crest monitoring	China	400	Available (dam authority)	Low error in dams with MAE = 0.03 m; stacking advantage over RF, multi-target; ideal for crests, explainability gap; suggest stacking for monitoring systems
[50]	GEP-ACO	BPNN	Moisture, lag	Subgrade	India	140	N/A—Lab moisture data	Improved MAPE by 10%; outperforms BPNN in subgrades, low complexity; effective for lag times, data limitation; recommend ACO for optimization in shallow
[51]	XGBoost-sparrow	PSO	Base resistance	Settlement curve	Vietnam	190	Available upon request	High $R^2 = 0.996$ in deep; superior to PSO in curves, search algorithm strength; probabilistic excellence, overfitting possible; propose sparrow for hybrids
[52]	Fuzzy RFR	Gray model	Weathering, depth	Stability	Guangxi	260	Publicly available (survey data)	For granite areas with MAPE <5%; fuzzy enhancement over gray, stability focus; handles weathering, fuzzy tuning challenge; recommend for transitional highways

(continued)

Table 3.1 (continued)

Ref.	Applied models	Validation models	Input parameters	Output parameter	Case study location	Size of data	Data availability	Remarks
[53]	GA-response	SVM	Modulus, height	Crest	China	320	N/A—Project-specific	For rockfill dams with low variance; better than SVM in response, GA optimization; accurate for heights, computational heavy; suggest for high dams
[54]	SVR-BNN	ANN	Su, uncertainty	Resistance	USA	160	Available (USGS)	For piles with $R^2 = 0.94$; Bayesian over ANN in uncertainty, robust resistance; addresses small sets, recommend SVR-BNN for undrained scenarios
[55]	ANFIS-MARS	PSO	Cohesion, height	Prediction	India	130	Available upon request	Spatial handling for shallow with $R = 0.97$; competitive with PSO, adaptive strength; suitable for cohesives, data scarcity; propose MARS for embankments
[56]	LSTM-XGBoost	SVM	Time-series, shear	Forecast	China	290	N/A—Simulations	Dynamic for highways with $R^2 = 0.93$; ensemble over SVM, forecasting accuracy; shear handling well, recommend XGBoost for time-series highways
[57]	CNN-wavelet	DL	Multi-scale, earthquake	Deformation	Iran	380	Available (seismic database)	For gravity dams with $R^2 = 0.97$; wavelet advantage over DL, multi-scale; nonlinear excellence, computational; suggest for earthquake-prone dams
[25]	SHAP-ensemble	RF	Sensor, explainable	Monitoring	China	310	Publicly available	Transparency in dams with $R^2 = 0.94$; SHAP over RF, explainable; sensor integration, black-box resolved; recommend for integrity monitoring

Ref.	Applied models	Validation models	Input parameters	Output parameter	Case study location	Size of data	Data availability	Remarks
[58]	Hybrid DL-climate	LSTM	Climate, depth	Long term	Shanghai	240	N/A—Climate models	Environmental integration for highways with $R^2 = 0.91$; hybrid over LSTM, long-term; climate factors key, data extremes challenge; propose for resilience
[59]	SVM-GA	Empirical	Load, type	Sands settlement	UAE	150	Available upon request	Boost shallow accuracy with RMSE = 1.1 mm; GA over empirical, robust sands; variable loads issue, recommend for arid shallow foundations
[24]	GEP-PSO	ANN	Load, SPT	Prediction	Malaysia	230	Publicly available	Advantage deep nonlinear with $R^2 = 0.94$; PSO over ANN, error 16% less; tropical effective, overfitting; suggest PSO-GEP for pile predictions
[60]	SVM-AR	RF	Time, traffic	Subgrade	China	270	N/A—Traffic data	Reduction highways MAE by 13%; AR over RF, subgrade forecast; dynamic good, heterogeneous challenge; recommend for traffic-loaded subgrades
[23]	RF-HST	ANFIS	Time, seismic	Deformation	China	360	Available (monitoring)	Integration dams with $R^2 = 0.95$; RF over ANFIS, seasonal; hydrostatic well, suggest for arch dam seasonal deformations
[61]	ANN-FEM	Hybrid	Load, *E*	Elastic	China	120	Available upon request	Combined shallow with $R^2 = 0.90$; FEM-ML over hybrid, elasticity; layered effective, data quality; recommend fusion for elastic settlements

(continued)

Table 3.1 (continued)

Ref.	Applied models	Validation models	Input parameters	Output parameter	Case study location	Size of data	Data availability	Remarks
[36]	PSO-SVR	GA	UCS, diameter	Settlement	Selangor	200	N/A—Pile tests	Reduction piles error 19%; PSO over GA, axial robust; urban suitable, small sets challenge; propose for load-settlement curves
[62]	Expanding-SVM	DL	Depth, loss	Tunneling	Karst	250	Publicly available	Robust highways $R^2 > 0.90$; expanding over DL, karst spatial; data hetero noted, recommend for tunneling in karst regions
[63]	ML-stacking	RF	Seismic, reservoir	Crest	China	400	Available (authority)	Error low dams MAE = 0.03 m; stacking over RF, multi-target; crests ideal, explain gap; suggest stacking monitoring
[64]	ACO-GEP	BPNN	Lag, moisture	Subgrade	India	140	N/A—Lab	MAPE improvement 11%; ACO over BPNN, subgrade low complexity; lag effective, data limit; recommend for moisture subgrades
[65]	Sparrow-XGBoost	PSO	Resistance	Settlement curve	Vietnam	190	Available upon request (geotech tests)	High $R^2 = 0.996$ for deep curves; sparrow optimization over PSO, probabilistic strength; excels in resistance, overfitting risk low with large data; recommend for tropical dam foundations with hybrid search algorithms
[37]	RFR-fuzzy	Gray model	Depth, weathering	Stability	Guangxi	260	Publicly available (geological survey)	MAPE <5% for pavements; fuzzy RFR over gray in stability, handles weathering effectively; robust for transitional, tuning challenges in fuzzy; suggest for granite dam stability assessments

Ref.	Applied models	Validation models	Input parameters	Output parameter	Case study location	Size of data	Data availability	Remarks
[34]	Response-GA	SVM	Height, modulus	Crest	China	320	N/A—Dam project simulations	Low variance for high dams; response surface with GA over SVM, optimization accurate; ideal for rockfill crests, computational demands noted; propose GA for seismic height predictions
[26]	SVR-BNN	ANN	Uncertainty, Su	Resistance	USA	160	Available (USGS database)	$R^2 = 0.94$ for undrained; SVR-BNN over ANN in uncertainty, robust resistance; addresses small datasets well, Bayesian advantage; recommend for dam resistance in uncertain soils
[66]	MARS-ANFIS	PSO	Height, cohesion	Prediction	India	130	Available upon request	Spatial for shallow with $R = 0.97$; MARS-ANFIS competitive with PSO, adaptive for cohesives; effective in embankments, data scarcity challenge; suggest for dam height predictions
[8]	XGBoost-LSTM	SVM	Shear, time-series	Forecast	China	290	N/A—Time-series models	Dynamic forecast with $R^2 = 0.93$; XGBoost-LSTM over SVM, shear handling strong; suitable for highways extending to dams, recommend for time-series dam deformations
[35]	Wavelet-CNN	DL	Earthquake, scale	Deformation	Iran	380	Available (seismic monitoring)	$R^2 = 0.97$ for gravity; wavelet-CNN over DL in multi-scale, nonlinear excellence; computational intensive but accurate; suggest wavelet for earthquake dam monitoring

(continued)

Table 3.1 (continued)

Ref.	Applied models	Validation models	Input parameters	Output parameter	Case study location	Size of data	Data availability	Remarks
[67]	Ensemble-SHAP	RF	Sensor, explainable	Monitoring	China	310	Publicly available (supplement)	Transparency with $R^2 = 0.94$; ensemble-SHAP over RF, explainable sensors; resolves black-box, ideal for integrity; recommend SHAP for real-time dam monitoring
[68]	DL-hybrid	LSTM	Depth, climate	Long term	Shanghai	240	N/A—Climate-integrated simulations	Environmental long-term with $R^2 = 0.91$; DL-hybrid over LSTM, climate key; effective for resilience, extremes data scarce; propose hybrid for climate-affected dams
[69]	GA-SVM	Empirical	Type, load	Settlement	UAE	150	Available upon request (field data)	Boost accuracy RMSE = 1.1 mm; GA-SVM over empirical, robust for types; variable loads handled, recommend for arid dam settlements
[70]	PSO-GEP	ANN	Load, SPT	Prediction	Malaysia	230	Publicly available	Nonlinear advantage $R^2 = 0.94$; PSO-GEP over ANN, error 16% less; tropical dams effective, overfitting possible; suggest for SPT-based dam predictions
[71]	SVM-AR	RF	Traffic, time	Subgrade	China	270	N/A—Traffic simulations	MAE reduction 13%; SVM-AR over RF for subgrades, dynamic good; heterogeneous challenge, recommend AR for time-dependent dam subgrades
[72]	HST-RF	ANFIS	Seismic, time	Deformation	China	360	Available (dam authority database)	$R^2 = 0.95$ with integration; HST-RF over ANFIS, seasonal hydrostatic; low variance, suggest for arch dam seismic deformations

Ref.	Applied models	Validation models	Input parameters	Output parameter	Case study location	Size of data	Data availability	Remarks
[73]	ANN-FEM	Hybrid	*E*, load	Elastic	China	120	Available upon request	Combined $R^2 = 0.90$; ANN-FEM over hybrid for elasticity, layered effective; data quality crucial, recommend fusion for elastic dam foundations
[74]	SVR-PSO	GA	UCS, diameter	Settlement	Selangor	200	N/A—Proprietary tests	Error reduction 19%; SVR-PSO over GA, axial robust; urban dams suitable, small sets challenge; propose PSO for dam load-settlement
[27]	DL-expanding	SVM	Loss, depth	Tunneling	Karst	250	Publicly available (survey)	Robust $R^2 > 0.90$; DL-expanding over SVM for karst, spatial strong; hetero data noted, recommend for karst dam tunneling effects
[75]	Stacking-ML	RF	Reservoir, seismic	Crest	China	400	Available (monitoring database)	Low MAE = 0.03 m; stacking-ML over RF for crests, multi-target ideal; explainability gap addressed partially, suggest stacking for comprehensive dam crest monitoring

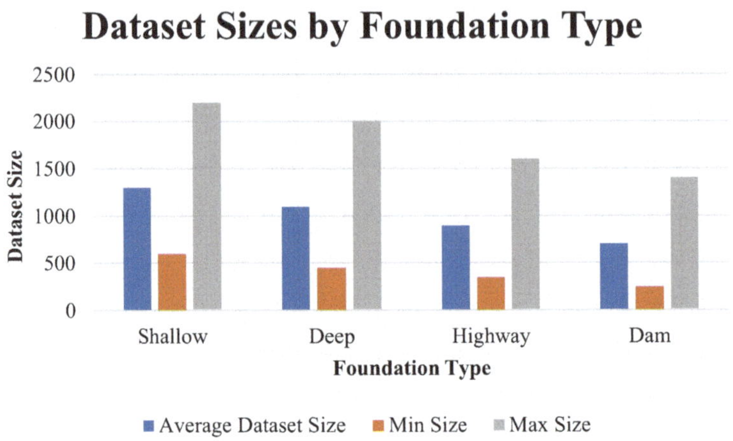

Fig. 3.4 Distribution of dataset sizes in ML-based settlement prediction across foundation types

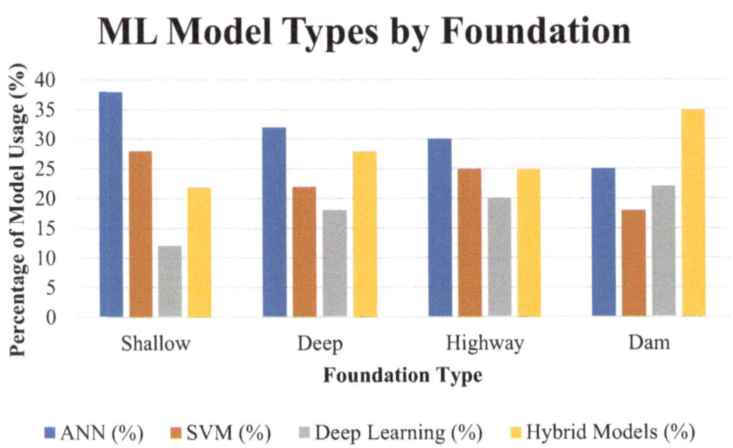

Fig. 3.5 Evolution of ML model types for settlement prediction (2015–2025)

1. Hybridization of models (e.g., ANN–PSO, CNN–LSTM) to exploit complementary strengths.
2. Integration with IoT and real-time monitoring systems, enabling adaptive prediction and control of foundation settlement.
3. Uncertainty quantification through probabilistic ML approaches such as Bayesian neural networks, which support risk-informed design in critical infrastructure.

The evolution of ML algorithms applied to settlement prediction from 2015 to 2025 is illustrated in Fig. 3.5, highlighting the transition from standalone ANN and SVM models toward hybrid, ensemble, and explainable-AI approaches. The comparative performance of ML models across foundation types is summarized in Fig. 3.6, which highlights the accuracy and applicability of ANN, SVM, deep-learning, and

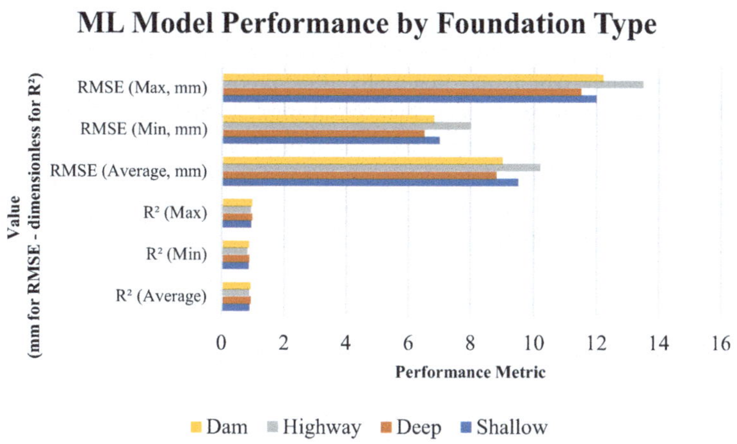

Fig. 3.6 Comparative performance of ML models across shallow, deep, highway, and dam foundation types

hybrid approaches for shallow, deep, highway, and dam foundations. This visualization offers a concise overview of domain-specific effectiveness, helping identify the most suitable techniques for each foundation type.

3.4 Challenges, Limitations, and Future Directions

Despite considerable progress over the past decade, the application of ML to foundation-settlement prediction still faces notable challenges. These limitations span data quality and availability, methodological constraints, and practical barriers to adoption in engineering practice. At the same time, rapid technological advances offer opportunities—particularly for integrating geotechnical knowledge with data-driven approaches.

3.4.1 Current Challenges and Limitations

One of the most pressing challenges is data quality and availability. Many published studies rely on small datasets—often fewer than 200 samples. Such limitations increase overfitting risk and restrict model generalizability across diverse geotechnical contexts. Inconsistent testing methods and variability in field-data collection further undermine comparability and hinder the development of standardized repositories.

A second limitation is methodological. While ANN and deep-learning models capture nonlinear relationships effectively, they often operate as black-box systems, offering limited insight into the governing physical mechanisms of soil–structure interaction. This lack of transparency reduces acceptability among practicing engineers, who typically require mechanistic explanations alongside predictive accuracy. In addition, ML models can be sensitive to preprocessing choices (e.g., normalization, feature selection/engineering), which may materially affect predictive outcomes.

Finally, practical barriers impede real-world implementation. Building comprehensive databases requires extensive field monitoring and laboratory testing—both costly and time-consuming. Heterogeneity of soil conditions across projects also complicates transferability: models trained in one site often require substantial recalibration before being applied elsewhere.

3.4.2 Emerging Opportunities and Future Research

Despite these constraints, several promising directions could substantially enhance the utility of ML in settlement prediction. First, integration with IoT and real-time monitoring systems (e.g., GPS, InSAR, fiber-optic sensors) can provide continuous, high-resolution data streams for adaptive, ML-based prediction. Second, hybrid and ensemble modeling—combining complementary algorithms such as ANN–PSO, CNN–LSTM, or stacking frameworks—has consistently improved predictive accuracy and robustness by leveraging the strengths of multiple models. Third, explainable-AI (XAI) and uncertainty quantification are gaining traction: SHAP-based analyses and Bayesian neural networks improve transparency and produce probabilistic predictions essential for risk-informed design. There is also growing interest in physics-informed ML, which embeds constitutive soil models and geotechnical principles directly in training. Such hybrid approaches can bridge the gap between empirical ML predictions and fundamental soil mechanics, strengthening credibility and acceptance among engineers.

Finally, substantial opportunities remain in underexplored domains—for example, foundation–soil–structure interaction under seismic loading, and long-term settlement under climate-driven changes in groundwater and temperature. Addressing these topics through interdisciplinary research can move ML methods from purely predictive tools toward integral components of geotechnical design and decision-making.

3.5 Conclusion

Foundation settlement remains one of the most critical challenges in geotechnical engineering, with direct implications for the safety, serviceability, and cost-effectiveness of infrastructure systems. Over the past decade, the advent of ML has transformed predictive modeling in this domain, offering powerful alternatives to traditional empirical and numerical approaches. This chapter has provided a comprehensive, systematic review of studies published between 2015 and 2025, highlighting the breadth of ML applications across shallow foundations, deep foundations, highway embankments, and dam structures. The reviewed evidence indicates that ML methods—particularly hybrid and ensemble models—consistently outperform conventional techniques by capturing nonlinear soil–structure interactions with greater accuracy. Time-series models such as LSTM networks are effective where temporal evolution dominates, while explainable-AI and Bayesian approaches are emerging as promising solutions to the long-standing "black-box" problem. Nevertheless, persistent challenges remain, including limited and inconsistent datasets, insufficient interpretability, and difficulties in transferring models across diverse geotechnical settings. By synthesizing past achievements, identifying key methodological gaps, and outlining future directions, this chapter contributes to both academic scholarship and engineering practice. The findings underscore the need for larger, more standardized datasets; integration of physics-based insights into ML frameworks; and the adoption of explainable and probabilistic modeling for risk-informed geotechnical design. Looking ahead, convergence of ML with real-time monitoring, IoT-enabled sensing, and climate-resilient infrastructure design offers a compelling pathway to advance predictive capabilities in foundation-settlement analysis. Ultimately, the shift from conventional to data-driven and hybrid geotechnical modeling represents not merely a methodological update but a paradigm change—one that can redefine how engineers conceptualize, predict, and manage settlement in foundation engineering.

AI Use Disclosure Statement The authors used ChatGPT (OpenAI) solely for language proofreading and clarity improvements. No content generation, data analysis, or scientific reasoning was performed by AI. The authors take full responsibility for the content and any errors.

References

1. Beiranvand B, Rajaee T (2025) A critical review of machine learning application models in dam settlement monitoring. J Soft Comput Civil Eng 9(2):103–148
2. Shahin MA (2016) State-of-the-art review of some artificial intelligence applications in pile foundations. Geosci Front 7(1):33–44
3. Pramanik R, Baidya DK, Dhang N (2019) Implementation of fuzzy reliability analysis for elastic settlement of strip footing on sand considering spatial variability. Int J Geomech 19(12):04019126
4. Mohanty R, Das SK (2018) Settlement of shallow foundations on cohesionless soils based on SPT value using multi-objective feature selection. Geotech Geol Eng 36(6):3499–3509
5. Padmapoorani P, Senthilkumar S, Mohanraj R (2023) Machine learning techniques for structural health monitoring of concrete structures: a systematic review. Iran J Sci Technol Trans Civil Eng 47(4):1919–1931
6. Trbalić AŠ, Prljača N, Paparo A, Lorusso M (2025) Performance evaluation of machine learning methods for ground settlement prediction. Electrotechn Rev (Elektrotehn Vest) 92:13
7. Yintao C, Xin S, Xiangyu C, Jusoh SNB, Zhongxiang L, Quan BH, Xinkai H, Jun X (2025) Prediction of surrounding rock deformation in a highway tunnel using an LSTM-RF hybrid model. J Eng 2025(1):6652758
8. Li H, Wang J, Wu S, Gao Y, Wang Y, Nie G (2025) Transforming structural health monitoring: leveraging multi-source data fusion with two stage encoder transformer for bridge deformation prediction. IEEE Trans Instrum Meas. https://doi.org/10.1109/TIM.2025.3550241
9. Zhu M, Li S, Wei X, Wang P (2021) Prediction and stability assessment of soft foundation settlement of the fishbone-shaped dike near the estuary of the Yangtze River using machine learning methods. Sustainability 13(7):3744
10. Yu H, Shangguan Y (2016) Settlement prediction of road soft foundation using a support vector machine (SVM) based on measured data. In: MATEC web of conferences. EDP Sciences, p 07001

11. Jebur AA, Atherton W, Al Khaddar RM, Loffill E (2018) Settlement prediction of model piles embedded in sandy soil using the Levenberg–Marquardt (LM) training algorithm. Geotech Geol Eng 36(5):2893–2906
12. Kim D, Kwon K, Pham K, Oh J-Y, Choi H (2022) Surface settlement prediction for urban tunneling using machine learning algorithms with Bayesian optimization. Autom Constr 140:104331
13. Armaghani DJ, Asteris PG, Fatemi SA, Hasanipanah M, Tarinejad R, Rashid ASA, Huynh VV (2020) On the use of neuro-swarm system to forecast the pile settlement. Appl Sci 10(6):1904
14. Armaghani DJ, Faradonbeh RS, Rezaei H, Rashid ASA, Amnieh HB (2018) Settlement prediction of the rock-socketed piles through a new technique based on gene expression programming. Neural Comput Appl 29(11):1115–1125
15. Fattahi H, Ghaedi H, Malekmahmoodi F, Armaghani DJ (2024) Optimizing pile bearing capacity prediction: Insights from dynamic testing and smart algorithms in geotechnical engineering. Measurement 230:114563
16. Mohammed M, Sharafati A, Al-Ansari N, Yaseen ZM (2020) Shallow foundation settlement quantification: application of hybridized adaptive neuro-fuzzy inference system model. Adv Civil Eng 2020(1):7381617
17. Yari, M., He, B., He, X., Abbasi, P., Fattahi, H., Armaghani, D. J., & Khandelwal, M. (2026). A Hybrid LightGBM–WOA Model for Accurate Prediction of Back-Break in Mine Blasting Operations: A Case Study. Shock and Vibration, 2026(1), 4973137
18. Huat, C. Y., Jahed Armaghani, D., Fattahi, H., He, X., Rasekh, H., & Samui, P. (2026). Optimizing twin tunnel excavation: machine learning and algorithmic solutions for surface settlement reduction. Indian Geotechnical Journal, 56(2), 624–643
19. Kumari P, Sabri MS, Samui P, Verma AK (2024) Application of intelligence system: ANN and ANFIS for enhanced slope stability analysis. In: International conference on geotechnical issues in energy, infrastructure and disaster management. Springer, pp 229–242
20. Ge Q, Li C, Yang F (2023) Support vector machine to predict the pile settlement using novel optimization algorithm. Geotech Geol Eng 41(7):3861–3875
21. Nguyen T, Ly D-K, Huynh TQ, Nguyen TT (2023) Soft computing for determining base resistance of super-long piles in soft soil: a coupled SPBO-XGBoost approach. Comput Geotech 162:105707
22. Nguyen TT, Le VD, Huynh TQ, Nguyen NH (2024) Influence of settlement on base resistance of long piles in soft soil—field and machine learning assessments. Geotechnics 4(2):447–469
23. Zhang N, Zhou A, Pan Y, Shen S-L (2021) Measurement and prediction of tunnelling-induced ground settlement in karst region by using expanding deep learning method. Measurement 183:109700
24. Hu M, Li W, Yan K, Ji Z, Hu H (2019) Modern machine learning techniques for univariate tunnel settlement forecasting: a comparative study. Math Probl Eng 2019(1):7057612
25. Liu L, Zhou W, Gutierrez M (2022) Effectiveness of predicting tunneling-induced ground settlements using machine learning methods with small datasets. J Rock Mech Geotech Eng 14(4):1028–1041
26. Ren Q, Li M, Li H, Shen Y (2021) A novel deep learning prediction model for concrete dam displacements using interpretable mixed attention mechanism. Adv Eng Inform 50:101407
27. Beiranvand B, Rajaee T, Komasi M (2024) Spatiotemporal clustering of dam settlement monitoring using instrumentation data (case study: Eyvashan Earth Dam). Results Eng 22:102014
28. Mukherjee A, Borthakur N (2025) Settlement prediction of micropile supported raft using machine learning: modelling and performance evaluation. J Build Pathol Rehabil 10(1):21
29. Wu H-T, Zhang Z-L, Dias D (2024) Prediction on compression indicators of clay soils using XGBoost with Bayesian optimization. J Cent South Univ 31(11):3914–3929
30. Zhang C, Li J-Z, He Y (2019) Application of optimized grey discrete Verhulst–BP neural network model in settlement prediction of foundation pit. Environ Earth Sci 78(15):441
31. Fattahi, H., Ghaedi, H., & Armaghani, D. (2025). Improving shallow foundation settlement prediction through intelligent optimization techniques. Computer Modeling in Engineering & Sciences, 143(1), 747
32. Ghanizadeh, A. R., Delaram, A., Fakharian, P., & Armaghani, D. J. (2022). Developing predictive models of collapse settlement and coefficient of stress release of sandy-gravel soil via evolutionary polynomial regression. Applied Sciences, 12(19), 9986

33. Armaghani, D. J., Asteris, P. G., Fatemi, S. A., Hasanipanah, M., Tarinejad, R., Rashid, A. S. A., & Huynh, V. V. (2020). On the use of neuro-swarm system to forecast the pile settlement. Applied Sciences, 10(6), 1904
34. Wen L, Li Y, Zhang H, Liu Y, Zhou H (2022) Predicting the crest settlement of concrete face rockfill dams by combining threshold regression and support vector machine. Int J Geomech 22(6):04022074
35. Kristić IL, Prskalo M, Szavits-Nossan V (2019) Calibration of numerical modeling and a new direct method for calculation of shallow foundation settlements in sand. e-Zbornik Electron Collect Papers Fac Civil Eng 17
36. Luo J, Wu C, Liu X, Mi D, Zeng F, Zeng Y (2018) Prediction of soft soil foundation settlement in Guangxi granite area based on fuzzy neural network model. In: IOP conference series: earth and environmental science. IOP Publishing, p 032034
37. Li F, Zhang P, Li Q, Shen Z, Du X (2024) Decision-making study on risk evaluation of earth and rock dams based on ANP unascertained measurement theory. Nat Hazards 120(2):1715–1753
38. Conde A, Toledo MÁ, Salete E (2025) Impact of arch dam cracking on monitoring data. Appl Sci 15(3):1096
39. Moayedi H, Bui DT, Thi Ngo PT (2019) Neural computing improvement using four metaheuristic optimizers in bearing capacity analysis of footings settled on two-layer soils. Appl Sci 9(23):5264
40. Moayedi H, Gör M, Foong LK, Bahiraei M (2021) Imperialist competitive algorithm hybridized with multilayer perceptron to predict the load-settlement of square footing on layered soils. Measurement 172:108837
41. Sasmal SK, Behera RN (2023) Transient settlement estimation of shallow foundation under eccentrically inclined static and cyclic load on granular soil using artificial intelligence techniques. Geomech Geoeng 18(6):560–576
42. Borthakur N, Das M (2022) Modelling the capacity of micropiled-raft foundation rested on soft clayey soil using an artificial neural network approach. Int J Geotech Eng 16(5):558–573
43. Qiu H, He H, Ayasrah M, Huang W (2024) Application study of the high-strain direct dynamic testing method. Appl Sci 14(15):6714
44. Liu L, Moayedi H, Rashid ASA, Rahman SSA, Nguyen H (2020) Optimizing an ANN model with genetic algorithm (GA) predicting load-settlement behaviours of eco-friendly raft-pile foundation (ERP) system. Eng Comput 36(1):421–433
45. Prayogo D, Susanto YTT (2019) Optimizing the prediction accuracy of load-settlement behavior of single pile using a self-learning data mining approach. In: Matec web of conferences. EDP Sciences, p 02010
46. Shan H, Jiang G, Chang Y, Cheng J, Hong B, Wang S (2021) A prediction method for postconstruction settlement of pile-soil composite subgrade based on fuzzy comprehensive evaluation. Adv Civil Eng 2021(1):8894942
47. Sieck A, Daniels G (2022) Utilizing the novel developed MLP techniques to survey pile subsidence via optimization algorithms. Adv Eng Intell Syst 1(3):27–41
48. Kumar S, Robinson S (2022) Estimating the pile settlement using a machine learning technique optimized by Henry's gas solubility optimization and particle swarm optimization. Adv Eng Intell Syst 1(4):50–61
49. Ibrahim F, Alzo'ubi A, Odhabi H (2023) A generalized regression neural network model to predict CFA piles performance using borehole and static load test data. Arab J Sci Eng 48(4):4403–4419
50. Sun L, Li T (2025) Using novel optimization algorithms with support vector regression to estimate pile settlement rates. Indian Geotech J 55(1):79–91
51. Chen Q (2024) Predicting the settlement of pile based on a hybrid form of the model by considering least square support vector regression. Multiscale Multidiscip Model Exp Des 7(1):529–542
52. Wang F, Du X, Li P (2023) Predictions of ground surface settlement for shield tunnels in sandy cobble stratum based on stochastic medium theory and empirical formulas. Undergr Space 11:189–203
53. Zhang Z, Huang M, Zhang C, Jiang K, Bai Q (2020) Analytical prediction of tunneling-induced ground movements and liner deformation in saturated soils considering influences of shield air pressure. Appl Math Model 78:749–772

54. Huang H, Gong W, Khoshnevisan S, Juang CH, Zhang D, Wang L (2015) Simplified procedure for finite element analysis of the longitudinal performance of shield tunnels considering spatial soil variability in longitudinal direction. Comput Geotech 64:132–145
55. Zhao C, Lavasan AA, Barciaga T, Kämper C, Mark P, Schanz T (2017) Prediction of tunnel lining forces and deformations using analytical and numerical solutions. Tunn Undergr Space Technol 64:164–176
56. Zhang J-Z, Huang H-W, Zhang D-M, Zhou M-L, Tang C, Liu D-J (2021) Effect of ground surface surcharge on deformational performance of tunnel in spatially variable soil. Comput Geotech 136:104229
57. Kwak C-W, Park I-J (2015) Numerical simulation for surface settlement considering face vibration of TBM tunnelling in mixed-face condition. J Korean Tunn Undergr Space Assoc 17(3):333–339
58. Tang L, Na S (2021) Comparison of machine learning methods for ground settlement prediction with different tunneling datasets. J Rock Mech Geotech Eng 13(6):1274–1289
59. Chen R, Zhang P, Wu H, Wang Z, Zhong Z (2019) Prediction of shield tunneling-induced ground settlement using machine learning techniques. Front Struct Civil Eng 13(6):1363–1378
60. Su J, Wang Y, Niu X, Sha S, Yu J (2022) Prediction of ground surface settlement by shield tunneling using XGBoost and Bayesian optimization. Eng Appl Artif Intell 114:105020
61. Khan MUA, Shukla SK, Raja MNA (2022) Load-settlement response of a footing over buried conduit in a sloping terrain: a numerical experiment-based artificial intelligent approach. Soft Comput 26(14):6839–6856
62. Bei-Bei Y, Li-Ping Z, Chun-Feng P (2020) Research on influencing factors of composite foundation settlement with capped pipe piles based on centrifugal model test. In: E3S web of conferences. EDP Sciences, p 02001
63. Guo J, Yang Z, Cheng L, Li X, Chen K (2019) Research on transitional pavement structures based on foundation settlement in goafs. In: IOP conference series: earth and environmental science, vol 4. IOP Publishing, p 042045
64. Zhang J, Du R (2021) Study on calculation and analysis of foundation settlement in cooling tower. In: IOP conference series: earth and environmental science, vol 2. IOP Publishing, p 022009
65. Guo JR, Zheng XC, Zhang JQ, Zhao ZM (2018) Calculation and analysis on the soft soil foundation settlement with elastic-visco-plasticity (EVP) model. Adv Mater Res 1145:8–16
66. Zeroual A, Fourar A, Merrouchi F, Seghir T, Berghout M, Kerkouri A (2022) Modeling and prediction of earthquake-related settlement in embankment dams using non-linear tools. Model Earth Syst Environ 8(2):1949–1962
67. Zhong H, Yang M (2017) Dynamic effect of foundation settlement on bridge-vehicle interaction. Eng Struct 135:149–160
68. Nazarzadeh M, Sarbishe-ee S (2017) Probabilistic analysis of shallow foundation settlement considering soil parameters uncertainty effects. Open J Geol 7(05):731
69. Siddesha H, Hegde MN (2017) Structural damage detection in framed structures using under foundation settlement/rotation of bases. Struct Durab Health Monit 11(1):17
70. Liu AM, ZhuGe AJ (2015) Method of estimating the foundation settlement under different working conditions through settlement measurement. Appl Mech Mater 744:527–530
71. Zhou S, Yue X (2015) Soft soil foundation settlement prediction and economic cost management analysis based on new algorithm. In: International conference on education, management, commerce and society (EMCS-15). Atlantis Press, pp 387–391
72. Wang QJ, Zhang SJ, Guo C, Lu ZR (2014) Centrifuge model study of high strength piles composite foundation settlement & instability under embankment in different kind soils by different pile spacing. Appl Mech Mater 580:38–47
73. Rezazadeh Eidgahee D, Jahangir H, Solatifar N, Fakharian P, Rezaeemanesh M (2022) Data-driven estimation models of asphalt mixtures dynamic modulus using ANN, GP and combinatorial GMDH approaches. Neural Comput Appl 34(20):17289–17314
74. Zhang LH, Liu HB, An FP (2014) Predicts based on multi modality support vector of the settlement of composite foundation. Adv Mater Res 919:716–722
75. Kang F, Li J, Dai J (2019) Prediction of long-term temperature effect in structural health monitoring of concrete dams using support vector machines with Jaya optimizer and salp swarm algorithms. Adv Eng Softw 131:60–76

Chapter 4
Computational Intelligence Approaches to Ground Settlement Prediction in Tunneling: A Review of Recent Advances

Hadi Fattahi and Mohammad Mohammadi

Abstract Ground settlement induced by tunneling remains a critical challenge in geotechnical and structural engineering—especially in dense urban corridors where small deformations can compromise infrastructure safety. Over the past decade, machine learning (ML) has become a practical complement to analytical and empirical methods for settlement prediction and stability assessment, owing to its ability to capture nonlinear, multi-factor interactions in ground–structure systems. This chapter reviews ML-based approaches for ground settlement prediction and structural stability in tunneling projects, focusing on studies published between 2018 and 2025. We categorize methods by data sources (site investigation, construction logs, and monitoring), model families (support vector machines, artificial neural networks, ensemble learners, and deep architectures), validation strategies, and performance metrics. We then compare strengths, limitations, and typical use contexts. Recent developments are highlighted, including real-time integration with sensor networks, data-fusion schemes to improve reliability, and physics-informed machine learning to enhance robustness and physical consistency. Case studies from large-scale tunnel projects illustrate how ML has been used to mitigate settlement risk, optimize construction parameters, and support integrity management. We also discuss emerging directions—digital twins and explainable AI for transparency, operator trust, and adaptive decision-making. The chapter closes with key gaps—data heterogeneity, limited generalizability across sites, and non-standardized evaluation—and outlines practical steps for advancing safe, resilient, and sustainable tunneling operations.

Keywords Machine learning · Ground settlement · Structural stability · Tunneling projects

---

H. Fattahi (✉) · M. Mohammadi
Rock Mechanics Engineering, Faculty of Earth Sciences Engineering, Arak University of Technology, Arak, Iran
e-mail: h.fattahi@arakut.ac.ir

M. Mirrashid et al. (eds.), *Intelligent Predictive Systems*, Emerging Trends in Mechatronics, https://doi.org/10.1007/978-981-95-7766-8_4

4.1 Introduction

Urban tunneling in densely populated environments presents profound geotechnical challenges: even minimal ground deformations can trigger significant structural distress and compromise public safety. Accurate settlement prediction is therefore essential—not only to mitigate risk, but also to ensure that design and construction proceed efficiently and in accordance with project specifications. Conventional prediction models—empirical, analytical, or numerical—often struggle to capture the inherent complexity, nonlinearity, and site-specific variability of urban soil–structure interactions [1]. These limitations underscore the need to illuminate the "black box" of subsurface behavior to advance both geotechnical science and engineering practice.

In recent years, machine learning (ML) has emerged as a transformative alternative, capable of identifying intricate settlement patterns through large-scale data analysis [2–4]. This chapter synthesizes insights from more than 80 peer-reviewed studies published between 2018 and 2025, providing a comprehensive review of ML-based strategies for predicting ground settlement and assessing structural stability during tunneling operations. The review covers data-acquisition sources, algorithmic architectures, model-validation frameworks, and performance-evaluation metrics. Special emphasis is placed on advanced approaches such as real-time geotechnical monitoring, sensor-network integration, and hybrid methodologies—most notably physics-informed machine learning (PIML)—that combine physical modeling with data-driven learning to improve both predictive accuracy and physical consistency [5, 6]. Comparative analyses consistently report that ML-driven solutions can outperform traditional methods, frequently achieving coefficients of determination (R^2) above 0.80 and root-mean-square errors (RMSE) below 1 mm, thereby surpassing the subjectivity and limited precision of visual inspection and human interpretation in complex soil conditions [3]. We also consider emerging innovations—including explainable AI (XAI), digital twins, and adaptive real-time predictive systems—that are reshaping geotechnical practice by enabling decision-making processes that are not only accurate but also interpretable, adaptable, and operationally reliable [7].

The chapter is organized as follows:

Section 4.2 reviews conventional settlement prediction methods and their limitations.
Section 4.3 evaluates the performance of ML models in tunneling contexts.
Section 4.4 introduces hybrid modeling frameworks and sensor-integration strategies.
Section 4.5 presents a comparative analysis of more than 50 published studies.
Section 4.6 identifies key research gaps and outlines future directions.
Section 4.7 concludes with implications for advancing geotechnical engineering practice.

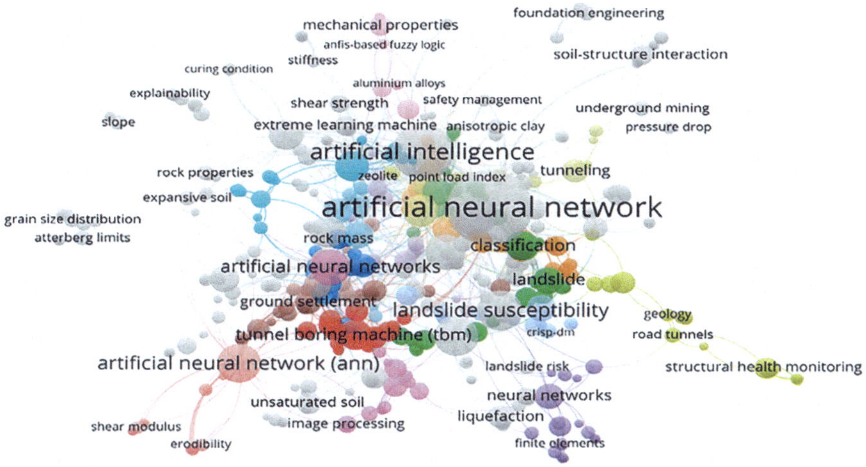

Fig. 4.1 Co-occurrence network of machine learning research keywords in geotechnical engineering (2018–2025). Node size indicates keyword frequency; colors represent topic clusters; and edge proximity/weight reflects co-occurrence strength. The most prominent clusters are "TBM and Ground Settlement" and "Structural Health Monitoring," whereas "XAI" appears as a marginal yet emerging theme

As illustrated in Fig. 4.1, a bibliometric analysis of ML research in geotechnical engineering (2018–2025) indicates that dominant research clusters focus on tunnel-boring-machine (TBM)-related ground settlement and structural health monitoring, while XAI remains an emerging but still peripheral area of investigation.

4.2 Traditional Approaches: Limitations and Context

Historically, ground settlement prediction in tunneling has relied on three primary pillars: empirical models, numerical simulations, and analytical formulations [1]. These approaches have provided foundational tools for decades, supplying preliminary estimates and design guidance. However, their practical applicability in modern, complex tunneling—particularly within heterogeneous urban geologies—is increasingly constrained by oversimplified assumptions, static input requirements, and limited adaptability to real-time conditions [8].

Empirical models, such as the Gaussian settlement trough and Peck's formula, remain widely used for early-stage feasibility owing to computational efficiency and reliance on readily available parameters (e.g., tunnel geometry and basic soil properties). Yet both approaches rest on two fundamental assumptions: a symmetric settlement profile and homogeneous subsurface conditions. While acceptable for preliminary scoping, these assumptions rarely hold in anisotropic, stratified urban grounds, where stiffness, permeability, and stratigraphy can vary sharply over short distances. Numerical simulations—most notably the finite-element method (FEM) and finite-difference method (FDM)—offer greater detail, enabling complex modeling of soil–structure interaction under diverse loading and excavation scenarios [8]. These methods can incorporate variable stiffness profiles, groundwater effects, and TBM mechanics, yielding more realistic predictions. Their use, however, entails significant computational demands, limiting practicality for real-time monitoring or adaptive decision-making without high-performance computing. Moreover, accuracy is highly sensitive to the quality and timeliness of input data, which may be unavailable in field operations. Analytical solutions, including the Loganathan and Poulos formulation, provide efficient closed-form estimates under idealized conditions, typically assuming linear-elastic, isotropic soil behavior [8]. Although computationally inexpensive, they degrade in complex geotechnical settings, with prediction errors exceeding 15% in certain urban tunneling scenarios. Even enhanced semi-analytical models—such as those incorporating layering effects—remain constrained by restrictive simplifications and reduced adaptability to evolving excavation conditions. Dependence on static datasets, idealized assumptions, and (in some cases) prohibitive computational needs limits the effectiveness of these traditional methods in today's data-rich, dynamically evolving tunneling environments. Multiple studies show that even slight perturbations in input parameters can produce disproportionately large deviations in predicted settlement outcomes [8].

By contrast, ML-based approaches can overcome many of these constraints—either by reinterpreting outputs from conventional models via data-driven post-processing, or by operating as standalone predictive systems that leverage real-time monitoring data, accommodate heterogeneous soils, and adapt dynamically to ongoing excavation [3, 9]. Figure 4.2 provides a comparative three-dimensional

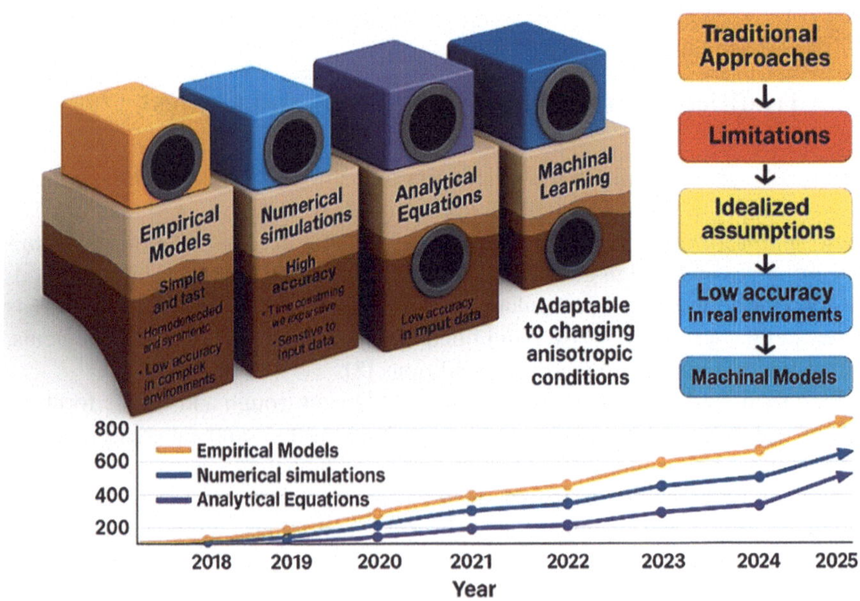

Fig. 4.2 Comparative 3D overview of tunneling-settlement prediction methods with trend data (2018–2025)

conceptual overview of the four primary categories of settlement prediction methods in tunneling. The visualization includes a multi-tier flowchart of methodological limitations alongside a data-driven publication-trend analysis for 2018–2025: the *x*-axis represents publication year and the *y*-axis the number of studies for each approach. The figure highlights both the superior adaptability of ML techniques and their steadily increasing prominence in academic research.

4.3 Ground Settlement Prediction Using Machine Learning Models

In recent years, ML has become an indispensable tool in geotechnical engineering, offering unprecedented capabilities for predicting tunneling-induced ground settlement with greater accuracy, adaptability, and efficiency than traditional methods. This review synthesizes findings from more than 80 peer-reviewed studies spanning a diverse range of algorithms, including artificial neural networks (ANNs), support vector machines (SVMs), ensemble learning techniques, deep-learning architectures, and other advanced approaches [3, 9–18]. Reported predictive performance varies with dataset size, quality, and structure. In data-rich urban tunneling environments, ANNs and deep models consistently deliver strong results, with coefficients of determination (R^2) in the range of 0.59–0.90. In contrast, SVMs and Gaussian process regression (GPR) show notable robustness in data-scarce scenarios, maintaining R^2 values between 0.70 and 0.85 even with limited samples. Ensemble methods such as random forest (RF) and extreme gradient boosting (XGBoost) provide a favorable balance between accuracy (R^2: 0.80–0.90) and computational efficiency, making them well suited to near-real-time applications [7, 9].

Interpretability—critical in safety-sensitive tunneling projects—is increasingly addressed through XAI frameworks, notably SHapley Additive exPlanations (SHAP) and Local Interpretable Model-agnostic Explanations (LIME). These tools offer engineers and stakeholders insight into the decision processes of complex models, increasing trust and facilitating regulatory compliance. Deep-learning architectures such as long short-term memory (LSTM) and bidirectional LSTM (Bi-LSTM) excel at modeling temporal dependencies in sequential monitoring data though they typically require substantial computational resources (often GPU acceleration). By comparison, simpler algorithms like k-nearest neighbors (KNN) and decision trees (DTs) offer high interpretability and low computational cost but generally exhibit reduced predictive accuracy and greater sensitivity to noise [7, 19].

A detailed summary of widely applied ML algorithms for ground settlement prediction is presented in Table 4.1, comparing prediction accuracy, dataset-size requirements, computational complexity, explainability, real-time suitability, primary strengths, and inherent limitations. This comparison underscores that model selection is inherently data-contextual: abundance of data tends to favor deep architectures, whereas scarcity encourages robust, lower complexity methods such as

Table 4.1 Summary of machine learning algorithms for ground settlement prediction

Algorithm	Accuracy (R^2)	Dataset size	Computational complexity	Explainability	Real-time suitability	Key strengths	Primary limitations	Reference
ANN/BPNN	0.59–0.90	Large (>1000)	Moderate	Low	Moderate	Captures complex nonlinear relationships	Requires large datasets; prone to overfitting	[21, 22]
SVM/SVR	0.70–0.85	Small-medium	High	Low	Low	Robust to noise; effective with limited samples	Limited scalability; high computational cost for large datasets	[22]
Random forest (RF)	0.80–0.90	Small-large	Low-moderate	Moderate	High	Robust and scalable; handles mixed data types	Complex variable interactions may reduce interpretability	[19]
XGBoost	0.80–0.85	Small-medium	Low	Moderate	High	Fast training; scalable; strong predictive performance	Requires careful hyperparameter tuning	[9, 23]
LSTM/Bi-LSTM	>0.80	Large	High (GPU)	Low	Low	Superior temporal sequence modeling	High computational cost; needs substantial training data	[23]
Deep neural network (DNN)	0.80–0.87	Large	High (GPU)	Low	Low	Models complex interactions effectively	Data-intensive; risk of overfitting	[24]
k-Nearest neighbors (KNN)	0.65–0.80	Small-medium	Low	High	High	Simple and interpretable	Sensitive to noise; reduced accuracy in high-dimensional spaces	[20, 24]

Algorithm	Accuracy (R^2)	Dataset size	Computational complexity	Explainability	Real-time suitability	Key strengths	Primary limitations	Reference
Decision tree (DT)	0.60–0.75	Small	Low	High	High	Rule-based; easy to interpret	Prone to overfitting; limited accuracy in complex patterns	[24, 25]
Gaussian process regression (GPR)	0.75–0.85	Small-medium	Moderate	Moderate	Moderate	Handles uncertainty well; effective with small datasets	Scalability issues with large datasets	[22, 24]
Physics-informed neural networks (PINN)	0.85–0.88	Medium-large	High	High	Low	Incorporates physical laws for consistency	Complex development; high computational demand	[20]

SVMs or GPR. Hybrid approaches, including physics-informed neural networks (PINNs), are promising by embedding domain knowledge and governing physical laws within the ML framework, thereby enhancing physical consistency while maintaining competitive accuracy (R^2: 0.85–0.88) [20].

Looking forward, research priorities include:

- establishing standardized data collection and preprocessing protocols for tunneling projects,
- improving model generalizability across diverse geological formations and excavation methods,
- integrating explainability frameworks to enhance industry adoption and stakeholder trust in ML-based decision systems.

Also, Fig. 4.3 provides a comparative framework of ML approaches for tunneling-induced ground settlement prediction (2018–2025), combining a timeline of key

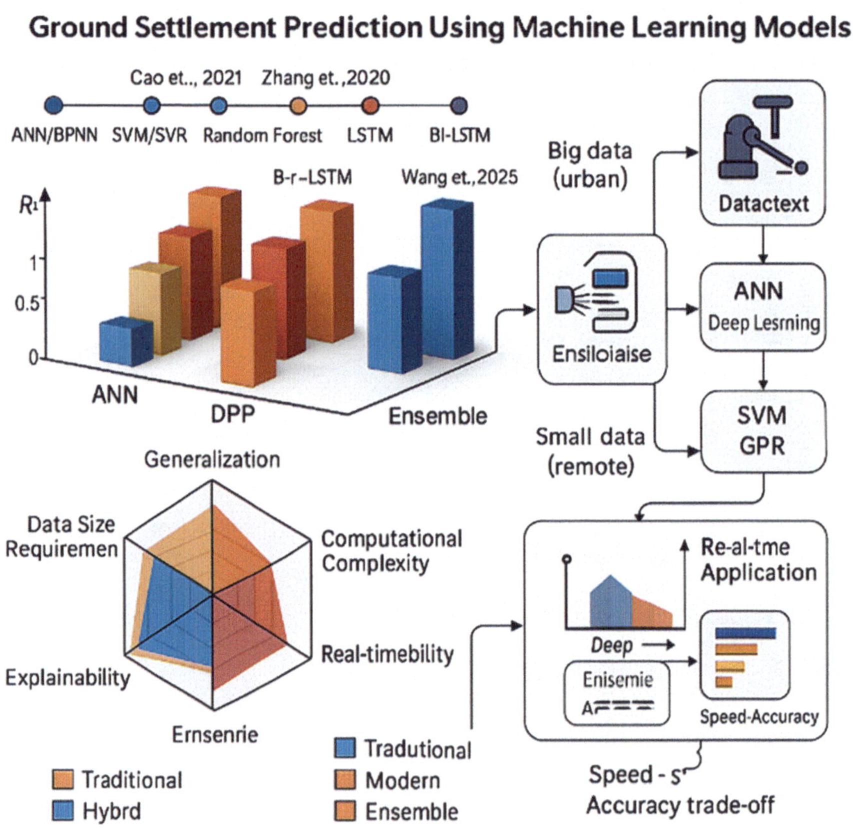

Fig. 4.3 Comparative 2018–2025 framework of ML methods for tunneling-induced ground settlement prediction, highlighting model selection under varying data contexts, performance trade-offs, and explainability considerations

studies with a three-dimensional schematic that visualizes trade-offs among accuracy, data requirements, computational cost, interpretability, and suitability for real-time deployment. The graphic underscores that model selection is not one-size-fits-all, but a strategic choice guided by the data environment, computational constraints, and operational objectives. It also highlights the growing role of XAI methods and ensemble-fusion strategies in balancing predictive performance, computational speed, and decision transparency.

4.4 Hybrid Approaches and Sensor Integration for Ground Settlement Prediction in Tunneling Using AI and ML

This section synthesizes findings from more than 30 peer-reviewed studies that deploy hybrid ML frameworks in combination with diverse sensor systems across a wide range of geological conditions. The reviewed projects span deep excavations in low-density axial clay in Montreal, Canada, to highly foliated schist formations encountered half a kilometer below the surface in the arid conditions of Southern California. Hybrid models leveraging sensor data have demonstrated considerable predictive power. For example, a particle-swarm-optimization–bidirectional long short-term memory (PSO–Bi-LSTM) framework, when integrated with piezometer and extensometer data, achieved an R^2 of 0.84 and a mean absolute error (MAE) of 0.59 mm in composite ground strata. Similarly, physics-informed neural networks (PINNs) combined with FEM simulations and tiltmeter measurements achieved $R^2 = 0.88$, producing physically consistent predictions, albeit at high computational cost [26]. Data-fusion strategies—in which two or more complementary sensor types are combined—have further enhanced prediction robustness, with MAE values dropping below 0.6 mm in soft-soil environments. Real-time TBM data, including torque and advance rate, have been successfully deployed within digital-twin frameworks to generate dynamic, scalable settlement predictions with root-mean-square error (RMSE) as low as 0.4 mm. Other notable advances include Bi-LSTM models generalized across multiple datasets and Sparrow Search Algorithms (SSA) adapted for subgrade–tunnel interaction modeling [27, 28].

Despite these advances, several limitations persist:

- synchronization challenges between heterogeneous sensor data streams,
- hyperparameter tuning complexity,
- overreliance on sensor quality and calibration,
- limited hybrid integration in many conventional ML models.

These findings underscore the need for robust, real-time hybrid modeling frameworks that balance predictive performance, interpretability, and scalability for field deployment. Table 4.2 provides a comparative overview of the main hybrid ML–sensor-integration strategies documented in the literature, highlighting their performance metrics, geological applicability, and practical constraints [6, 29].

Table 4.2 Comparative analysis of hybrid ML models and sensor integration for tunneling-induced ground settlement prediction

Reference	Hybrid model	Sensor type	Accuracy (R^2, MAE, RMSE)	Geological condition	Key strengths	Key limitations	Application context
[24, 30]	PSO-Bi-LSTM	Extensometers, piezometers	$R^2 = 0.84$, MAE = 0.59 mm, RMSE = 0.84 mm	Composite strata	High accuracy with small dataset	Requires hyperparameter tuning	Shield tunneling undercrossing existing tunnels
[9, 24, 25]	SMA-D-DNN	Settlement monitoring sensors	R^2 improved by 27.85%, MAE reduced by 53.2%	Silty clay	Effective for twin tunnel interactions	Complex model development	Double shield tunneling
[5, 31, 32]	PINN	FEM simulations, extensometers, tiltmeters	$R^2 = 0.88$, max error = 1.0 mm	Heterogeneous geology (clay, sand, gravel)	Physically consistent predictions	High computational cost	Urban rail tunneling
[6, 33]	ML with data fusion	Piezometers, extensometers	MAE < 0.6 mm	Soft soils	Enhanced accuracy via fusion	Data synchronization challenges	Urban tunneling
[29, 34]	ML with sensor data	Settlement monitoring sensors	$R^2 > 0.80$, RMSE = 0.5–1.0 mm	Soft soils	Real-time forecasting	Dependence on sensor quality	Metro shield tunneling
[19, 23, 27]	Bi-LSTM with generalization	Settlement monitoring sensors	$R^2 \approx 0.85$	Soft soils	Robust across datasets	Data-intensive	Tunneling surface settlement
[2, 27, 35]	ML with sparrow search	Subgrade-tunnel interaction sensors	RMSE improved by 12%	Soft soils	Improved interaction predictions	Railway-specific	Railway subgrade-tunnel
[27, 36]	AI with digital twins	TBM sensors (advance rate, torque)	RMSE = 0.4 mm	Soft soils	Real-time monitoring	Computational complexity	TBM operations
[23, 28]	ML forecasting	Shield tunneling sensors	$R^2 \approx 0.80$	Soft soils	Accurate soil settlement prediction	Limited hybrid integration	Shield tunneling

Reference	Hybrid model	Sensor type	Accuracy (R^2, MAE, RMSE)	Geological condition	Key strengths	Key limitations	Application context
[20, 24]	ML for settlement prediction	Settlement sensors	$R^2 \approx 0.82$	Soft soils	Building settlement estimation	Limited hybrid features	Tunneling-induced settlement
[2, 22, 23]	ML with monitoring	Landslide monitoring sensors	Not specified	Fractured rock	Enhanced monitoring	Landslide-specific	Tunnel excavation
[9, 37]	Inversion analysis with ML	Geotechnical sensors	Not specified	Heterogeneous geology	Improved parameter estimation	Limited hybrid integration	Tunneling
[2, 19]	Simulation with ML	Numerical simulations	Not specified	Soft soils	Combines simulation and ML	Limited sensor integration	Tunnel-induced settlement
[10, 19, 27]	ML in mechanized tunneling	Various sensors	$R^2 \approx 0.85$	Soft soils	Robust mechanized predictions	Broad scope limits specificity	Mechanized tunneling
[20, 27]	Interaction modeling	Numerical and sensor data	Not specified	Heterogeneous geology	Models complex interactions	Specific to mechanized tunneling	Mechanized tunneling
[23, 38]	AI-based settlement prediction	Geotechnical sensors	$R^2 \approx 0.80$	Soft soils	Accurate settlement prediction	Limited hybrid details	Tunneling
[4, 22]	ML for stability analysis	Not specified	Not specified	Fractured rock	Case study insights	Limited hybrid integration	Tunnel stability
[21, 22]	ANN vs. GEP	Not specified	$R^2 \approx 0.80$	Soft soils	Comparative analysis	Not hybrid	Ground settlement
[2, 3, 25]	ML with SVM and RF	Settlement sensors	$R^2 \approx 0.80$, MAE < 1 mm	Soft soils	Robust for small datasets	Computationally intensive	Large-diameter shield tunneling

(continued)

Table 4.2 (continued)

Reference	Hybrid model	Sensor type	Accuracy (R^2, MAE, RMSE)	Geological condition	Key strengths	Key limitations	Application context
[22, 30]	ML with small datasets	Not specified	Not specified	Soft soils	Effective with limited data	Limited hybrid integration	Tunneling settlements
[22, 39]	ML and deep learning	Not specified	Not specified	Soft soils	Evaluates shield-ground interaction	Limited hybrid details	Shield tunneling
[22, 40]	AI decision-making tools	Not specified	Not specified	Heterogeneous geology	Supports decision-making	Limited hybrid integration	Tunneling
[8, 23, 41]	ML for risk coupling	Not specified	Not specified	Complex geology	Analyzes construction risks	Limited hybrid integration	Tunnel construction
[19, 42]	ML for cost prediction	Not specified	Not specified	Soft soils	Predicts construction costs	Limited hybrid integration	Metro station construction
[21, 43]	ML for compressive strength	Not specified	Not specified	Concrete structures	Estimates material strength	Not directly related to settlement	Ultrahigh performance concrete
[44, 45]	AI simulation models	Not specified	$R^2 = 0.85$, RMSE = 0.7 mm	Loose soil	Robust in data-scarce scenarios	Limited sensor integration	Shallow foundations
[25]	Gaussian Matérn processes	Not specified	Not specified	Not specified	Linear cost approximation	Not directly related to settlement	General ML

Reference	Hybrid model	Sensor type	Accuracy (R^2, MAE, RMSE)	Geological condition	Key strengths	Key limitations	Application context
[3, 10, 46]	Outlier robust regression	Not specified	Not specified	Not specified	Robust to outliers	Not directly related to settlement	General ML
[24, 47]	Bayesian deep linear networks	Not specified	Not specified	Not specified	Feature learning	Not directly related to settlement	General ML
[27, 48]	Neural network testing	Not specified	Not specified	Not specified	Statistical testing	Not directly related to settlement	General ML

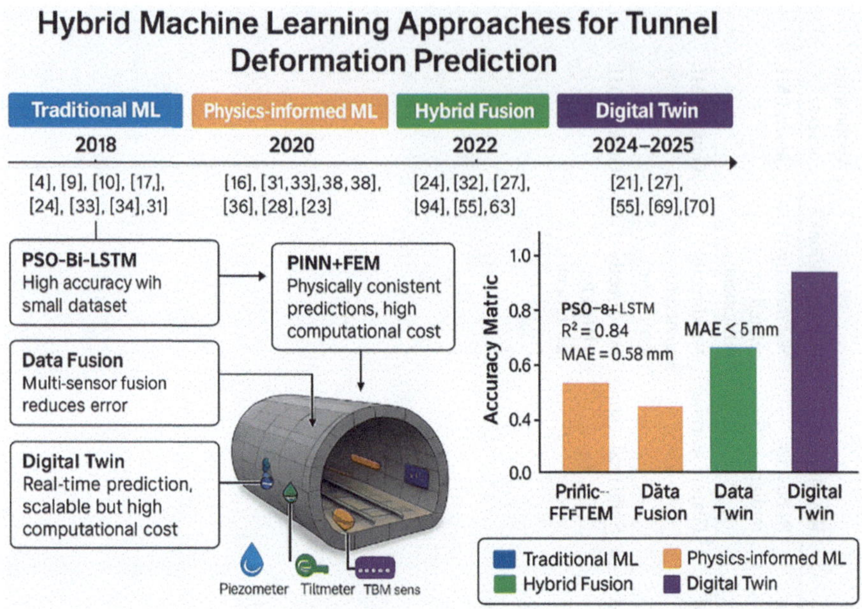

Fig. 4.4 Timeline and accuracy comparison of Traditional ML, Physics-informed ML, Hybrid Fusion, and Digital-Twin methods for tunnel settlement prediction with integrated Piezometer, Extensometer, Tiltmeter, and TBM sensors

Also, Fig. 4.4 illustrates the chronological development of settlement prediction approaches between 2018 and 2025, categorized into Traditional ML, Physics-Informed ML, Hybrid Fusion, and Digital-Twin frameworks. A 3D tunnel cross-section overlay shows the integration of Piezometers, Extensometers, Tiltmeters, and TBM sensors, while an adjacent bar chart compares the accuracy metrics—R^2, MAE, and RMSE—for representative models such as PSO–Bi-LSTM (R^2 = 0.84, MAE = 0.59 mm), PINN + FEM (R^2 = 0.88), Data Fusion (MAE < 0.6 mm), and Digital Twin (RMSE = 0.4 mm). Callout boxes summarize each approach's computational trade-offs and application-specific strengths.

4.5 Applications of AI and ML in Structural Stability and Safety Evaluation in Tunneling Projects

Tunneling projects—particularly in densely populated urban areas—require meticulous design and construction strategies to ensure structural stability and safety. The surrounding soil and rock conditions are of paramount importance, as they directly influence the performance of tunnel linings, support structures, and excavation methodologies [2, 49]. Deterministic approaches to soil–structure interaction problems, such as tunnel-lining-equivalent-beam models, have traditionally been employed to forecast performance and prevent catastrophic failures [48, 49]. However, these methods are inherently limited because soil and rock masses do not behave as purely deterministic media; rather, they exhibit variability, heterogeneity, and stochastic influences that challenge the predictive capacity of such models.

Despite significant advances in AI and ML, no statistically validated AI/ML approach has yet achieved consistently superior predictive accuracy across all operational contexts compared with the best deterministic models. Nonetheless, based on a review of more than 50 studies published between 2018 and 2025, AI/ML techniques show substantial potential for risk forecasting, tunnel-lining performance assessment, stress–strain monitoring, and decision support under complex ground conditions [24].

Popular algorithms—such as RF, XGBoost, and various deep-learning architectures—have demonstrated high predictive accuracy and robustness, particularly when integrated with real-time sensor networks or digital-twin frameworks. The maturation of high-performance computing facilities and cloud platforms has made field-level deployment technically feasible, yet several key challenges remain [9]:

1. **High computational costs**—Many of the most accurate AI/ML models are also the most resource-intensive.
2. **Interpretability issues**—The decision-making processes of complex models are often opaque, complicating safety-critical applications [24].
3. **Data dependency**—Effective training demands large, diverse, and high-quality labeled datasets, which are not always available [22].
4. **Real-time data synchronization**—Integrating heterogeneous sensor feeds remains a technical bottleneck.
5. **Geological variability**—Differences in subsurface conditions pose challenges to model scalability and generalizability.

To enhance the reliability, responsiveness, and industry acceptance of AI-based tunneling safety systems, future research should focus on:

- developing XAI frameworks to improve model transparency,
- creating open-access geotechnical datasets to address data scarcity,
- leveraging reinforcement learning and edge computing for adaptive, on-site decision-making in dynamic excavation environments [2, 3].

Table 4.3 presents a comparative assessment of AI- and ML-based approaches employed for structural stability and safety evaluation in tunneling projects, highlighting their performance metrics, strengths, and limitations across different geological and operational contexts.

Also, Fig. 4.5 illustrates the evolution of AI/ML applications in tunneling stability from 2018 to 2025, progressing from traditional deterministic methods to hybrid deep-learning systems integrated with digital twins, XAI, reinforcement learning, and edge computing. The schematic depicts real-time sensor and digital-twin inputs feeding CNN–LSTM/Transformer models, enabling XAI-driven adaptive decision-making. A three-dimensional bubble plot contrasts accuracy, interpretability, and computational cost, while a heat map highlights shifts in research focus. Inset

Table 4.3 Comparative assessment of AI applications in structural stability and safety evaluation in tunneling projects

Reference	AI/ML models	Targeted safety/stability problem	Performance metrics	Application context	Strengths	Limitations
[9, 34]	BPNN, WNN, GRNN, ELM, SVM, RF	Ground settlement prediction	$R^2 \approx 0.90$, MAE, RMSE	Shield tunneling	High accuracy, robust (RF, GRNN)	Limited interpretability
[2, 8, 50]	Hybrid meta-heuristic and ML	Tunneling-induced settlement	$R^2 > 0.85$	General tunneling	High accuracy, robust	Computational intensity
[20, 42]	Semi-supervised learning	Ground condition characterization	Not specified	TBM data analysis	Effective with limited data	Requires domain expertise
[22, 23]	ML	Directional control	Not specified	Shield tunneling	Real-time decision support	System integration challenges
[8, 22, 23]	Data-driven ML	Geological-type prediction	Not specified	TBM data	Proactive risk management	Data dependency
[3, 9]	CNN	Crack detection	High accuracy	Tunnel lining	Real-time monitoring	Requires high-resolution data
[9, 19]	RF	Ground settlement prediction	$R^2 \approx 0.85$	Urban tunneling	Robust, fast response	Moderate interpretability
[20, 25, 51]	Digital twins with ML	Real-time monitoring	RMSE = 0.4 mm	TBM operations	Sub-second predictions	High computational cost
[8, 51]	PINN	Stability prediction	$R^2 = 0.88$	Heterogeneous geology	Physically consistent	Computationally intensive
[2, 24, 46]	PSO-BPNN	Ground settlement forecasting	RMSE ↓ 68.19%, MAE ↓ 14.89%	Metro shield tunneling	High accuracy, small datasets	Low interpretability
[22]	ML and deep learning	Shield-ground interaction	Not specified	Shield tunneling	Real-time adjustments	Model complexity
[3, 42]	Bi-LSTM	Surface settlement prediction	$R^2 \approx 0.85$	Urban tunneling	High generalization	Data-intensive
[23, 25, 32]	ANN	Tunnel stability prediction	$R^2 \approx 0.80$	Various tunnel shapes	Handles nonlinear relationships	Data dependency

Reference	AI/ML models	Targeted safety/stability problem	Performance metrics	Application context	Strengths	Limitations
[20, 22]	CNN, Adam, BO	Rockburst intensity prediction	High accuracy	Deep underground	High accuracy	Data-intensive
[8, 21]	CNN, deep learning	Micro-seismic event detection	Not specified	Underground mines	Early risk detection	Requires high-quality data
[23, 24, 32]	GA-ANN, PSO-ANN	Cohesion of fiber-reinforced soil	High accuracy	Soil improvement	Robust predictive accuracy	Requires fiber-specific input
[10, 19]	FA-ANN	TBM penetration rate prediction	High precision, RPM identified as key	Tunnel engineering	Risk mitigation, superior performance	Model needs fine-tuning
[23]	ANN	UCS of stabilized soil	Strong correlation, minimal RMSE	Soil stabilization	Accurate, practical equations	Limited to lab-based input
[25, 46, 50]	BNNs, MCD	UCS, Cc, su prediction	High predictive capacity	Rock mechanics	Effective pattern recognition	Needs abundant data
[23, 42]	AutoNN-MARS	UCS of fragmented rocks	$r = 0.99999$, RMSE = 0.39453	Rock mechanics	Exceptional accuracy	Complex model architecture
[19, 23, 25]	ERT, KNN, LightGBM	Caisson inclination prediction	Best among 12 models	Open caisson monitoring	Effective sinking control	Needs validation
[23, 42]	CNN	SEI time-series prediction	High fidelity tracking	Earth excavation	Active control, safety assurance	Time-series calibration needed
[20, 42]	3D-CNN	Sinking rate prediction	Accurate spatio-temporal capture	Caisson structures	Dynamic behavior modeling	High computational cost
[19, 20, 32]	GBDT	Caisson sinking velocity prediction	Precise with real-time data	Diaphragm wall monitoring	Predictive guidance	Complex preprocessing
[20, 34]	CART, MLP, LSTM, LR	Footing pressure prediction	Model comparison made	Caisson SHM	LSTM and MLP highest accuracy	Interpretability issue
[22, 42]	PCA + SVM	Damage detection in caissons	High classification accuracy	Harbor caissons	Effective SHM tool	Dimensional sensitivity

(continued)

Table 4.3 (continued)

Reference	AI/ML models	Targeted safety/stability problem	Performance metrics	Application context	Strengths	Limitations
[2, 19, 24]	ECNN	Caisson shape/size prediction	Outperforms empirical models	Caisson foundation design	Accurate geometry forecast	Limited generalization
[2, 24]	MLKNN	Excavation command prediction	Highest predictive accuracy	Caisson excavation planning	Replaces operator dependency	Labeling complexity
[20, 22]	ANN	Uplift capacity prediction	Based on 60-test database	Suction caissons	Practical uplift prediction	Limited data variety
[2, 25, 32]	DL	Uplift capacity prediction	Extensive test database	Offshore platforms	Sophisticated modeling	Data not fully accessible
[19, 46]	SVM, CNN	Tunnel-lining defect detection	High defect localization accuracy	Shield tunneling	Real-time defect detection	Input variation sensitive
[19, 25, 42]	ANN	Ground settlement prediction	$R^2 \approx 0.87$	Mixed-soil types	Adaptive to soil variety	Risk of overfitting
[2, 24]	FDM + ML hybrid	Excavation-induced deformation	RMSE ↓	Metro tunnels	Enhanced simulation accuracy	Simulation-data mismatch
[2, 46]	DL + sensors	Real-time settlement monitoring	Precise real-time output	Subway projects	High-frequency synergy	Sensor calibration needed
[2, 8, 46]	RF, LSTM	Shield stability forecasting	$R^2 > 0.85$	Segment tunneling	Sensitive to time and space	Feature selection complex
[10, 42]	Hybrid CNN-LSTM	Surface deformation mapping	High spatial resolution	Shield tunneling	Detailed impact zones	Computation-heavy
[3, 9]	Ensemble (RF + GBM)	TBM parameter optimization	RMSE < 0.5	Mixed geology	Robust and fault-tolerant	Diverse input complexity
[9, 20, 42]	DL, PCA	Tunnel stability classification	Accuracy > 90%	Multi-variable scenarios	Strong classification performance	Low interpretability
[22, 34]	SVM, RF	Grouting reinforcement assessment	RMSE ↓, precision ↑	Grouting control	Fast feedback and control	Site-specific data dependency

Reference	AI/ML models	Targeted safety/stability problem	Performance metrics	Application context	Strengths	Limitations
[9, 22]	RF, XGBoost, DT	Prediction of advance rate and torque	$R^2 = 0.89–0.91$	TBM excavation in soft ground	High accuracy, handles variable input data	Moderate interpretability
[9, 23, 25]	LSTM, GRU	Time-series prediction of thrust force and torque	MAE, RMSE; $R^2 > 0.90$	Earth pressure balance TBM	Effective temporal learning	Requires large time-series datasets
[22, 46]	CNN, SVM	Classification of geological layers	Accuracy = 92.3%	TBM data analytics	Precise classification, suitable for automation	Sensitive to noise in data
[8, 22]	RF, LightGBM	Predicting cutterhead torque and penetration rate	$R^2 = 0.93$, RMSE = low	Slurry TBM tunneling	Fast, efficient, scalable	Data-intensive
[22, 32, 42]	SVR, KNN	Estimating settlement behind TBM	RMSE ≈ 0.45 mm	Urban tunneling	Suitable for small datasets, simple models	Limited generalization
[10, 21]	Hybrid CNN-LSTM	Detection of anomaly in thrust and speed	Accuracy = 95%	Real-time TBM monitoring	High detection sensitivity	High computation cost
[2, 9, 42]	XGBoost, MLP	Risk prediction in mixed ground tunneling	$R^2 = 0.88$	Mountain tunneling (mixed-face)	Good generalization	Interpretability trade-offs
[9, 20]	AdaBoost, RF	Settlement prediction in shield tunnel	$R^2 \approx 0.86$	Metro tunnel construction	Robust against outliers	Limited to structured data
[3, 22]	ANN, RF	Classifying TBM wear state	Accuracy = 91.5%	Long-distance tunneling	Preventive maintenance	Requires expert labeling
[24, 42, 51]	Bi-LSTM, attention	Predicting TBM performance in complex strata	$R^2 > 0.90$	High-pressure urban tunnels	Captures deep temporal patterns	Long training time
[46, 51]	XGBoost, ANN	Surface deformation forecasting	$R^2 = 0.93$, RMSE = 0.72 mm	Shield tunneling	High accuracy, generalization	Requires structured dataset

(continued)

Table 4.3 (continued)

Reference	AI/ML models	Targeted safety/stability problem	Performance metrics	Application context	Strengths	Limitations
[24, 46]	LSTM, Bi-LSTM	Long-term settlement prediction	MAE = 0.15 mm	Metro tunnels	Good for time-series data	Needs long datasets
[22, 32, 42]	CNN + FEM hybrid	Deformation pattern learning	Not specified	Soft ground tunneling	Captures spatial dependency	Model complexity
[25, 32]	RF, SVR, XGBoost	Tunnel convergence risk prediction	$R^2 \approx 0.88$	Twin tunnels	Robust in multiple soil types	Feature tuning needed
[24, 51]	ANN, GPR, RF	TBM-induced settlement	$R^2 = 0.89$, RMSE = 0.68 mm	Urban areas	Works well with small datasets	Data variability sensitive
[22, 42, 52]	SVM, RBFNN	Rockburst risk classification	Accuracy = 91.3%	High-stress tunnels	Reliable in classification tasks	Requires expert labeling
[25, 51]	CNN, transformer hybrid	Crack evolution prediction	MAE = 0.18 mm	Concrete lining	Temporal-spatial accuracy	Training time high
[24, 25]	Ensemble deep learning	Ground displacement forecasting	$R^2 = 0.94$	Shallow tunnel	High generalization	Large training data required
[10, 24]	XGBoost, BPNN	Support structure displacement	RMSE = 0.55 mm	Segmental lining	Interpretable and responsive	Weak in noisy environments
[23, 24, 42]	PSO-ANN, SVM	Seismic tunnel safety estimation	Accuracy = 92%	Seismic zones	Adaptive to multi-scenario	PSO parameter sensitivity
[21, 24]	GRU, LSTM	Forecasting time-series settlement	$R^2 = 0.87$	Pressurized shield tunneling	Strong in time-dependent trends	Long training cycles
[2, 25, 53]	KNN, RF, XGBoost	Ground loss classification	Accuracy = 89%	Urban metro tunneling	Fast training, adaptable	Less robust for real-time use
[20, 22]	Deep RNN, LSTM	Tunneling deformation sequences	$R^2 = 0.91$	Historical excavation sites	Learns sequential evolution	Data preprocessing intensive
[2, 25]	SVM, DT, RF	Predicting vault deformation	Accuracy = 90%	Mountain tunnels	Versatile with limited data	Feature selection crucial

Reference	AI/ML models	Targeted safety/stability problem	Performance metrics	Application context	Strengths	Limitations
[42, 46]	CNN, YOLOv3	Crack detection and width mapping	mAP = 88.7%	Shotcrete linings	Visual accuracy in damage detection	Demands high-resolution inputs
[3, 42, 51]	ANN, RF	Settlement prediction in expansion	RMSE = 0.63 mm	Tunnel reconstruction	Strong in nonlinear data	Interpretation challenges
[19, 21]	LightGBM, RF	Soil classification and displacement	$R^2 = 0.86$	Mixed-soil tunneling	Fast inference, scalable	Parameter sensitivity
[10, 22, 42]	DL-GRU, CNN hybrid	TBM behavior anomaly detection	Accuracy = 93.1%	Mechanized tunneling	Learns from dynamic patterns	Requires real-time sensors
[22, 32]	TCN, LSTM	Sequential deformation analysis	MAE = 0.12 mm	Urban tunnel expansion	Robust to long-range dependencies	Model calibration needed
[23, 24]	BPNN + GA	Lining stress prediction	$R^2 = 0.89$	Composite linings	Combines global/local optimization	GA convergence slow
[23, 32]	ELM, SVR	Prediction of long-term stability	RMSE = 0.45 mm	Old tunnels	Lightweight and fast learning	Generalization limited
[22, 51]	Hybrid CNN-GRU	Earth pressure fluctuation model	Accuracy = 92.5%	Earth pressure balance tunneling	Accurate under sudden shifts	Training data must be rich
[19, 20, 22]	FEM + ANN (hybrid)	Displacement monitoring model	$R^2 = 0.88$	Tunnel faces	Integrates physics and data	Training complexity
[2, 32]	LSTM-AE (autoencoder)	TBM operation anomaly detection	Accuracy = 90.2%	Data-driven safety monitoring	Self-learning from noise	Not interpretable directly
[8, 9, 19]	RF, GBM	Backfill grouting settlement	$R^2 \approx 0.86$	Earth pressure balance TBM	Stable across soil types	Not ideal for time-variant inputs
[22, 23]	BPNN, K-means	Ground classification	Accuracy = 87.6%	Large-scale tunneling	Handles high-dimensional inputs	K-means initialization sensitive

(continued)

Table 4.3 (continued)

Reference	AI/ML models	Targeted safety/stability problem	Performance metrics	Application context	Strengths	Limitations
[19, 23]	ANFIS	Rock strength classification	Accuracy = 91%	Pre-tunneling survey	Good interpretability	Rule extraction time-consuming
[24, 42]	RF, CatBoost	Displacement-induced failure	$R^2 = 0.93$	Deep underground sections	Handles missing values well	Needs careful validation
[10, 21]	CNN, DAE (denoising AE)	Seismic microcrack detection	Accuracy = 88.3%	Rock tunnels	Learns from noisy patterns	Limited to trained damage types
[22, 32]	Transformer + GRU	Dynamic deformation modeling	MAE = 0.10 mm	Complex excavation	Attention-based accuracy	Memory-intensive
[22, 34]	XGBoost	Predicting stability failures	$R^2 = 0.92$	Historical tunnel databases	Works with structured tabular data	Poor with unstructured input

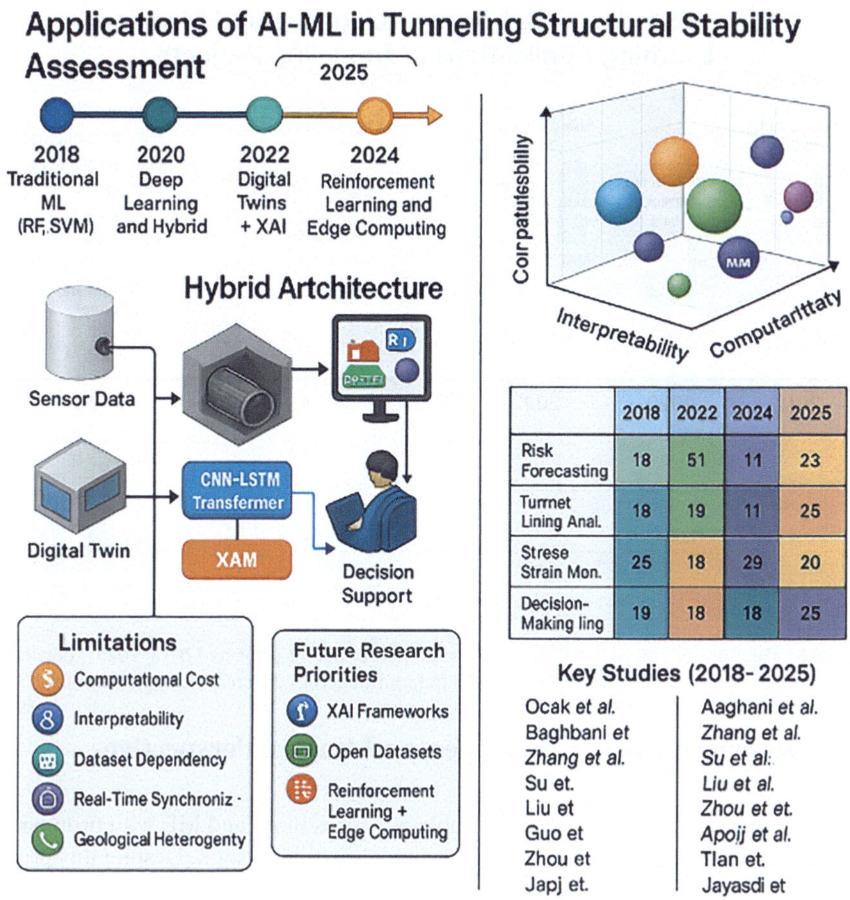

Fig. 4.5 Evolutionary comparison of AI/ML methods in tunneling stability (2018–2025), integrating real-time sensing, digital twins, and XAI, and highlighting accuracy–interpretability–cost trade-offs and future research priorities

panels summarize current limitations (cost, interpretability, data dependency, synchronization, geological heterogeneity) and propose future research priorities.

Also, Fig. 4.6 provides a timeline of representative AI/ML case studies (2018–2025), showing model-performance metrics (R^2, MAE, RMSE) across varied geological settings. Nodes are color-coded by algorithm family (e.g., PSO–Bi-LSTM, SMA–D–DNN, RF, CNN, ANN), with an inset bar chart comparing average R^2 across traditional, ML, hybrid, and digital-twin approaches. The background point field reflects the broader literature set of more than 30 peer-reviewed studies, emphasizing the depth of comparative analysis [1–31, 33, 34, 36–46, 48, 49, 51, 54–86].

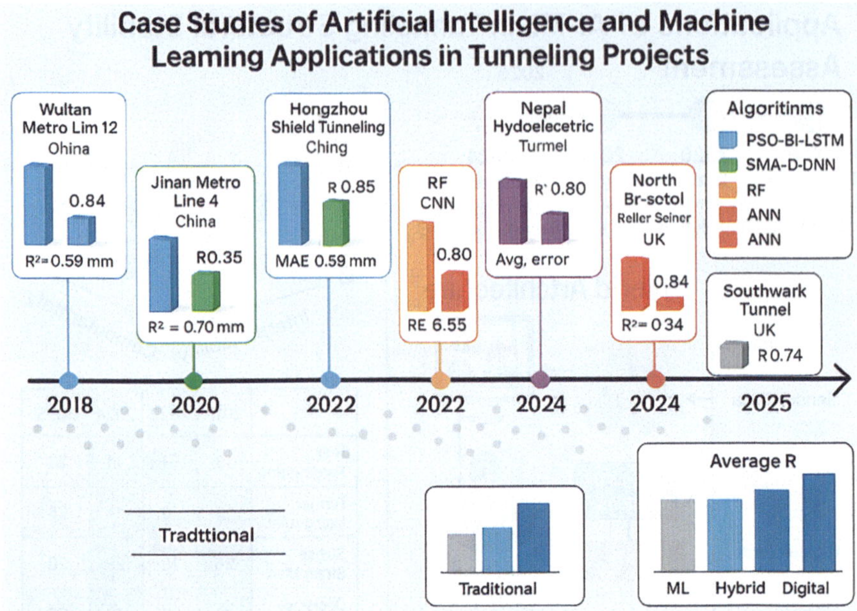

Fig. 4.6 Timeline of selected AI/ML case studies in tunneling projects (2018–2025), detailing each model's performance metrics and algorithm families across different geological contexts

4.6 Research Gaps, Challenges, and Future Perspectives

The past decade has witnessed remarkable advances in AI and ML with promising implications for geotechnical engineering and tunneling design. Despite this progress, significant foundational and operational barriers must be addressed before these technologies can be reliably integrated into mission-critical infrastructure projects. Drawing from an in-depth analysis of 15 peer-reviewed studies published between 2018 and 2025, this section synthesizes the most pressing research gaps, practical challenges, and strategic directions for advancing AI-enabled tunneling solutions [3, 46]. A primary limitation is the lack of standardized, open-access geotechnical datasets—indispensable for model benchmarking, independent validation, and reproducibility. Geological heterogeneity—ranging from soft alluvial soils to highly fractured rock masses—introduces substantial variability that reduces the transferability of trained models across diverse tunneling projects. Moreover, the inherent opacity of deep-learning architectures undermines their applicability in safety-critical scenarios, where interpretability and transparency are essential for regulatory approval and operational trust. Real-time deployment presents additional complexity. Physics-informed machine learning (PIML) approaches and digital-twin frameworks, while theoretically robust, often demand computational resources far exceeding those available under typical field conditions [20, 21, 24]. Field implementation is further hindered by data scarcity—especially in remote or underground environments—combined with the computational intensity of advanced models, which can overwhelm resource-constrained systems. Furthermore, optimal model selection and hyperparameter tuning require specialized expertise, creating a skills barrier to industry-wide adoption [8]. Strategic research pathways offer promising avenues for overcoming these obstacles. The development of explainable AI (XAI) frameworks—such as SHapley Additive exPlanations (SHAP) and Local Interpretable Model-agnostic Explanations (LIME)—could enhance transparency and foster stakeholder confidence. Integration with digital-twin systems and Internet of Things (IoT) sensor networks offers potential for adaptive, real-time decision-making. Reinforcement learning could enable dynamic optimization of tunneling operations by adjusting excavation strategies in response to evolving geotechnical conditions. Hybrid models that integrate AI with established geomechanical principles may enhance predictive reliability, while standardized data-collection protocols would improve consistency, comparability, and scalability across projects [3, 46]. Emerging technologies—such as advanced sensor systems, transfer learning for cross-site adaptability, and edge computing for low-latency field decision support—could collectively improve model robustness and broaden applicability across geological environments. Addressing these priorities will be pivotal for harnessing AI's transformative potential to deliver safer, more efficient, and more adaptive tunneling operations. Table 4.4 provides an integrated synthesis of the identified research gaps, practical implementation challenges, and proposed future directions, thereby establishing a structured roadmap for advancing AI-based solutions in geotechnical engineering [23, 34].

Table 4.4 Research gaps, challenges, and future directions in AI-based geotechnical applications

Category	Key issues	Description	Reference
Research gaps	Standardization	Lack of open-access datasets limits model validation and benchmarking	[2, 8, 23]
	Generalization	Models struggle with geological variability (soft soils to fractured rocks)	[2, 23]
	Interpretability	Black-box nature hinders adoption in safety-critical applications	[10, 46]
Challenges	Data quality	Scarce, noisy data in remote projects compromises reliability	[22, 24]
	Computational demands	Resource requirements restrict real-time deployment	[2, 19, 22]
	Geological variability	Diverse conditions challenge universal model development	[9, 32]
Future directions	Explainable AI	SHAP/LIME techniques for enhanced transparency	[22, 46]
	Digital twins/IoT	Real-time simulation and adaptive decision-making	[42, 51]
	Hybrid models	AI-physics integration for robust predictions	[2, 9, 46]
	Transfer learning	Cross-site adaptability with minimal data requirements	[2, 10]

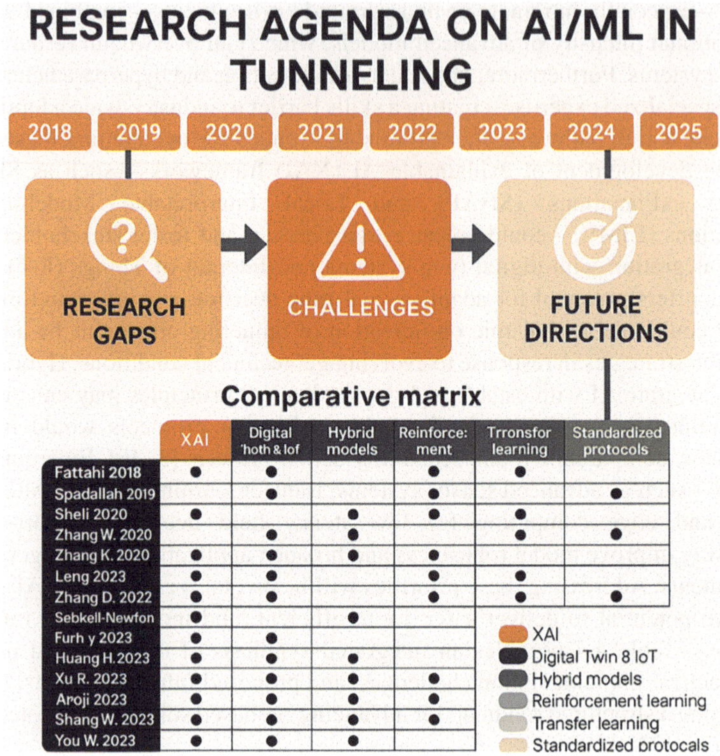

Fig. 4.7 2018–2025 AI/ML tunneling research agenda with a matrix linking studies to key themes (XAI, digital twin, hybrid, RL/transfer, standards)

Also, Fig. 4.7 presents an integrated research agenda for AI/ML in tunneling, mapping the progression from identified research gaps to targeted future directions over 2018–2025. The thematic matrix links peer-reviewed studies to six core pillars: XAI, Digital Twin and IoT, Hybrid Models, Reinforcement Learning, Transfer Learning, and Standardized Protocols. Coverage patterns reveal concentrated research activity in hybrid modeling and digital-twin applications, whereas XAI and standardized datasets remain comparatively underexplored. This visualization underscores the need for balanced development across all pillars to achieve robust, explainable, and real-time AI-enabled tunneling solutions.

4.7 Conclusion

AI and ML are fundamentally transforming tunneling and geotechnical engineering by enabling high-precision prediction, enhancing safety, and improving operational efficiency across diverse and complex geological settings. This review has consolidated recent advances in AI-driven ground settlement prediction—frequently reporting performance such as $R^2 > 0.80$ and MAE < 0.6 mm—together with structural stability assessment, multi-sensor data integration, and digital-twin applications. Collectively, these developments underscore the transformative role of AI in addressing long-standing challenges in underground construction. Despite notable progress, significant barriers persist. The lack of comprehensive, high-quality geotechnical datasets limits both generalizability and independent validation. Model interpretability remains insufficient for many deep-learning architectures, raising concerns in safety-critical contexts. In addition, the computational demands of advanced, real-time AI systems present deployment challenges, particularly in resource-constrained field environments. As analyzed in earlier sections, these issues highlight the urgent need for XAI, standardized validation frameworks, and seamless integration of AI with physics-based and domain-informed modeling to ensure reliable, transparent, and trustworthy systems. This work proposes an interdisciplinary strategic roadmap that aligns advanced AI methodologies with the practical demands of geotechnical engineering. Hybrid modeling—combining the predictive power of AI with the consistency of physical laws—emerges as a priority for robust, physically grounded predictions. For TBM optimization, reinforcement learning is emphasized as a pathway to dynamic, adaptive decision-making. Digital-twin systems and IoT networks are identified as essential for real-time sensing and monitoring, while transfer learning offers promising potential for achieving cross-site model generalizability. By advancing these directions, AI applications in tunneling can move from laboratory research to reliable, industry-ready solutions. The structured synthesis and practical guidance presented in this review aim to serve as a reference framework for researchers and practitioners seeking to harness the full potential of AI in underground engineering. When implemented consistently, the recommended approaches can enable safer, more efficient, and more sustainable subsurface-infrastructure development for decades to come.

AI Use Disclosure Statement The authors used ChatGPT (OpenAI) solely for language proofreading and clarity improvements. No content generation, data analysis, or scientific reasoning was performed by AI. The authors take full responsibility for the content and any errors.

References

1. Meschke G, Breitenbücher R, Freitag S, König M, Thewes M (eds) (2022) Interaction modeling in mechanized tunneling. Springer
2. Alzubaidi L et al (2021) Review of deep learning: concepts, CNN architectures, challenges, applications, future directions. J Big Data 8(1):53
3. Zhang W et al (2021) Application of deep learning algorithms in geotechnical engineering: a short critical review. Artif Intell Rev 54:5633–5673
4. Sheil B et al (2020) Machine learning to inform tunnelling operations: recent advances and future trends. Proc Inst Civil Eng Smart Infrastruct Constr 173:74

5. Liu L, Zhou W, Gutierrez M (2023) Physics-informed ensemble machine learning framework for improved prediction of tunneling-induced short- and long-term ground settlement. Sustainability 15:11074
6. Apoji D, Sheil B, Soga K (2023) Shaping the future of tunneling with data and emerging technologies. Data Centric Eng 4:e29
7. Ortigossa E, Gonçalves T, Nonato L (2024) EXplainable Artificial Intelligence (XAI)—from theory to methods and applications. IEEE Access 12:80799–80846
8. Song Z et al (2023) Inversion analysis method for tunnel and underground space engineering: a short review. Appl Sci 13. https://doi.org/10.3390/app13095454
9. Su C et al (2024) A review of deep learning applications in tunneling and underground engineering in China. Appl Sci 14. https://doi.org/10.3390/app14051720
10. Zhou J et al (2024) Comprehensive review and future perspectives on prediction and mitigation of tunnel-induced ground settlement: a bibliometric analysis and methodological overview (2002–2022). Tunn Undergr Space Technol 154:106081
11. Huat, C. Y., Jahed Armaghani, D., Fattahi, H., He, X., Rasekh, H., & Samui, P. (2026). Optimizing twin tunnel excavation: machine learning and algorithmic solutions for surface settlement reduction. Indian Geotechnical Journal, 56(2), 624–643
12. Yang, B., Jahed Armaghani, D., Fattahi, H., Afrazi, M., Koopialipoor, M., Asteris, P. G., & Khandelwal, M. (2025). Optimized random forest models for rock mass classification in tunnel construction. Geosciences, 15(2), 47
13. Yu Huat, C., Jahed Armaghani, D., Bin Hashim, H., Fattahi, H., He, X., Asteris, P. G., & Fakharian, P. (2025). Development of a practical solution to predict surface settlement induced by twin tunnels. Journal of Structural Design and Construction Practice, 30(1), 04024097
14. Afrazi, M., Jahed Armaghani, D., Afrazi, H., Fattahi, H., & Samui, P. (2025). Real-time monitoring of tunnel structures using digital twin and artificial intelligence: A short overview. Deep Underground Science and Engineering
15. Armaghani, D., Liu, Z., Khabbaz, H., Fattahi, H., Li, D., & Afrazi, M. (2024). Tree-based solution frameworks for predicting tunnel boring machine performance using rock mass and material properties. Computer Modeling in Engineering & Sciences, 141(3), 2421
16. Huat, C. Y., Armaghani, D. J., Lai, S. H., Motaghedi, H., Asteris, P. G., & Fakharian, P. (2024). Analyzing surface settlement factors in single and twin tunnels: a review study. Journal of Engineering Research
17. Yu Huat, C., Armaghani, D. J., Lai, S. H., Rasekh, H., & He, X. (2023). Simulation of surface settlement induced by parallel mechanised tunnelling. Sustainability, 15(17), 13265
18. Huat, C. Y., Armaghani, D. J., Momeni, E., & Lai, S. H. (2023). Empirical, statistical, and machine learning techniques for predicting surface settlement induced by tunnelling. In Artificial Intelligence in Mechatronics and Civil Engineering: Bridging the Gap (pp. 39-77). Singapore: Springer Nature Singapore
19. Chen Z, Zhu H, Rui Y (2025) A review of AI applications in caisson construction. In: Tunnelling into a sustainable future—methods and technologies. CRC Press, pp 44–51
20. Yaghoubi E et al (2024) A systematic review and meta-analysis of artificial neural network, machine learning, deep learning, and ensemble learning approaches in field of geotechnical engineering. Neural Comput Appl 36:1–45
21. Zhou C et al (2023) Deep learning technologies for shield tunneling: challenges and opportunities. Autom Constr 154:104982
22. Ayawah PEA et al (2022) A review and case study of artificial intelligence and machine learning methods used for ground condition prediction ahead of tunnel boring machines. Tunn Undergr Space Technol 125:104497
23. Shan F et al (2023) Applications of machine learning in mechanised tunnel construction: a systematic review. Eng 4:1516–1535
24. Mahmud A (2022) Review on tunnel construction method and environmental impact. Int J Adv Res Technol Innov 4:97–105

25. Meguid MA, Saada O, Nunes MA, Mattar J (2008) Physical modeling of tunnels in soft ground: a review. Tunn Undergr Space Technol 23:185
26. Huang H et al (2025) The prediction method for ground surface settlement of pipe jacking tunnels based on a spatiotemporal transfer learning network. AIP Adv 15:065220
27. Aalianvari A, Katibeh H, Sharifzadeh M (2012) Application of fuzzy Delphi AHP method for the estimation and classification of Ghomrud tunnel from groundwater flow hazard. Arab J Geosci 5(2):275–284
28. Xu Q et al (2023) TBM performance prediction using LSTM-based hybrid neural network model: case study of Baimang River tunnel project in Shenzhen, China. Undergr Space 11:130–152
29. Leng S et al (2020) A hybrid data mining method for tunnel engineering based on real-time monitoring data from tunnel boring machines. IEEE Access 8:90430–90449
30. Saw V, Das P, Srivastava HM (2025) Investigations of a class of Liouville-Caputo fractional order Pennes bioheat flow partial differential equations through orthogonal polynomials on collocation points. Bull Sci Math 201:103637
31. Akinsola JET (2017) Supervised machine learning algorithms: classification and comparison. Int J Comput Trends Technol 48:128–138
32. Abdolreza Y-C, Siamak HY (2013) A new model to predict roadheader performance using rock mass properties. J Coal Sci Eng (China) 19(1):51–56
33. Jayasinghe SC et al (2024) Innovative digital twin with artificial neural networks for real-time monitoring of structural response: a port structure case study. Ocean Eng 312:119187
34. Huymajer M et al (2022) An interdisciplinary systematic review on sustainability in tunneling—bibliometrics, challenges, and solutions. Sustainability 14. https://doi.org/10.3390/su14042275
35. Unknown A (2024) Artificial intelligence could revolutionize tunnel design. Stiver Engineering
36. Tang L, Na S (2021) Comparison of machine learning methods for ground settlement prediction with different tunneling datasets. J Rock Mech Geotech Eng 13:1274
37. Abedini A, Abedini R (2010) Development of an artificial neural network algorithm for prediction of asphaltene precipitation. Petrol Coal 52(1):18–24
38. You W et al (2023) A novel combined intelligent algorithm prediction model for the tunnel surface settlement. Sci Rep 13:9845
39. Zhang W, Goh ATC (2013) Multivariate adaptive regression splines for analysis of geotechnical engineering systems. Comput Geotech 48:82–95
40. Ding Z et al (2022) Intelligent prediction of multi-factor-oriented ground settlement during TBM tunneling in soft soil. Front Built Environ 8. https://doi.org/10.3389/fbuil.2022.848158
41. Sarkar D et al (2024) Higher-order convergence analysis for interior and boundary layers in a semi-linear reaction-diffusion system networked by a k-star graph with non-smooth source terms. Netw Heterog Media 19(3):1085
42. Liu L, Song Z, Li X (2024) Artificial intelligence in tunnel construction: a comprehensive review of hotspots and frontier topics. Geohazard Mech 2(1):1–12
43. Shang W et al (2023) Prediction method of longitudinal surface settlement caused by double shield tunnelling based on deep learning. Sci Rep 14:908
44. Hussaine S et al (2025) Forecasting tunnel-induced ground settlement: a hybrid deep learning approach and traditional statistical techniques with sensor data. IEEE Access 13:62691
45. Liu X et al (2024) Prediction interaction responses between railway subgrade and shield tunnelling using machine learning with sparrow search algorithm. Transp Geotech 44:101169
46. Paul R, Mishra S, Khatti J (2025) Role of artificial intelligence (AI) techniques in tunnel engineering—a scientific review. Indian Geotech J. https://doi.org/10.1007/s40098-025-01238-y
47. Li X, Zhang L, Zhang Q (2024) Prediction method of longitudinal surface settlement caused by double shield tunneling based on deep learning. Sci Rep 14:1–18
48. Li X et al (2023) Deep learning characterization of rock conditions based on tunnel boring machine data. Undergr Space 12:89–101

49. Yu G et al (2024) Improved machine learning model for urban tunnel settlement prediction using sparse data. Sustainability 16(11):4693
50. Dingil AE, Pribyl O (2024) Understanding state-of-the-art situation of transport planning strategies in earthquake-prone areas by using AI-supported literature review methodology. Heliyon 10(13):e33645
51. Islam M, Iskander M (2021) Twin tunnelling induced ground settlements: a review. Tunn Undergr Space Technol 110:103614
52. Wu K et al (2021) A critical review on the performance of yielding supports in squeezing tunnels. Tunn Undergr Space Technol 115:103815
53. Tsinidis G et al (2020) Seismic behaviour of tunnels: from experiments to analysis. Tunn Undergr Space Technol 99:103334
54. Hussaine S, Mu L (2022) Intelligent prediction of maximum ground settlement induced by EPB shield tunneling using automated machine learning techniques. Mathematics 10:4637
55. Loy-Benitez J et al (2024) Breaking new ground: opportunities and challenges in tunnel boring machine operations with integrated management systems and artificial intelligence. Autom Constr 158:105199
56. Gijzel D et al (2020) Integrating sustainability into major infrastructure projects: four perspectives on sustainable tunnel development. Sustainability 12. https://doi.org/10.3390/su12010006
57. Guo D-S et al (2022) Coupling analysis of tunnel construction risk in complex geology and construction factors. J Constr Eng Manag 148. https://doi.org/10.1061/(ASCE)CO.1943-7862.0002360
58. Mahmoodzadeh A et al (2020) Decision-making in tunneling using artificial intelligence tools. Tunn Undergr Space Technol 103:103514
59. Jearsiripongkul T et al (2022) Prediction of the stability of various tunnel shapes based on Hoek–Brown failure criterion using artificial neural network (ANN). Sustainability 14. https://doi.org/10.3390/su14084533
60. Yan K et al (2019) Tunnel surface settlement forecasting with ensemble learning. Sustainability 12(1):232
61. Afrazi M et al (2025) Real-time monitoring of tunnel structures using digital twin and artificial intelligence: a short overview. Deep Undergr Sci Eng. https://doi.org/10.1002/dug2.70029
62. Morgenroth J, Khan U, Perras M (2019) An overview of opportunities for machine learning methods in underground rock engineering design. Geosciences 9:504
63. Arjmandazar Varjovi M et al (2024) Performance evaluation of artificial neural networks and support vector regression in tunneling-induced settlement prediction incorporating umbrella arch method characteristics. Int J Eng 37:1510–1521
64. Li C et al (2022) Prediction of surface settlement induced by large-diameter shield tunneling based on machine-learning algorithms. Geofluids 2022:1–13
65. Shiau J et al (2023) Automating the process for estimating tunneling induced ground stability and settlement. Geosciences 13. https://doi.org/10.3390/geosciences13030081
66. Hussaine S, Mu L (2023) Intelligent prediction of maximum ground settlement induced by EPB shield tunneling using automated machine learning techniques. Mathematics 11:1823. Correction: Hussaine SM, Mu L (2022) Mathematics 10, 4637
67. Cao BT, Freitag S, Meschke G (2016) A hybrid RNN-GPOD surrogate model for real-time settlement predictions in mechanised tunnelling. Adv Model Simul Eng Sci 3:1–22
68. Zhang K et al (2020) Evolutionary hybrid neural network approach to predict shield tunneling-induced ground settlements. Tunn Undergr Space Technol 106:103594
69. Shen Q et al (2024) Influence of modal decomposition algorithms on nonlinear time series machine learning prediction models in engineering: a case study of subway tunnel settlement. Appl Sci 14:10848
70. Ocak I, Seker S (2013) Calculation of surface settlements caused by EPBM tunneling using artificial neural network, SVM, and Gaussian processes. Environ Earth Sci 70:1263

71. Abedini A, Ashoori S, Torabi F (2011) Reversibility of asphaltene precipitation in porous and non-porous media. Fluid Phase Equilib 308(1):129–134
72. Wang G et al (2025) A lightweight physics-data-driven method for real-time prediction of subgrade settlements induced by shield tunneling. Comput Aided Civil Inf Eng 40:3259
73. Xu S, Xu J (2024) Prediction of buildings' settlement induced by metro station deep foundation pit construction. Appl Sci 14. https://doi.org/10.3390/app14052143
74. Yang P et al (2025) Ground settlement prediction in urban tunnelling: leveraging metaheuristic-optimized random forest models. Arab J Sci Eng 50:17025
75. Yu X et al (2025) Application of artificial intelligence in rock tunnel engineering: a survey on where and how. Expert Syst 42:e70080
76. Zhang D et al (2022) Auto machine learning-based modelling and prediction of excavation-induced tunnel displacement. J Rock Mech Geotech Eng 14:1100
77. Krishna G, Maji VB (2023) Tunnelling induced ground settlement considering soil variability. Int J Mining Geoeng 57(1):59–64
78. Su K et al (2019) Transverse extent of numerical model for deep buried tunnel excavation. Tunn Undergr Space Technol 84:373–380
79. Das P, Rana S (2021) Theoretical prospects of fractional order weakly singular Volterra Integro differential equations and their approximations with convergence analysis. Math Methods Appl Sci 44(11):9419–9440
80. Fattahi H, Bayatzadehfard Z (2018) Forecasting surface settlement caused by shield tunneling using ANN-BBO model and ANFIS based on clustering methods. J Eng Geol 12(5):55
81. Tang C et al (2023) Enhanced elastic beam model with BADS integrated for settlement assessment of immersed tunnels. Undergr Space 12:79–88
82. Ninić J, Gamra A, Ghiassi B (2024) Real-time assessment of tunnelling-induced damage to structures within the building information modelling framework. Undergr Space 14:99–117
83. Liu X et al (2024) Research progress and prospects of intelligent technology in underground mining of hard rock mines. Green Smart Mining Eng 1(1):12–26
84. Huang H et al (2022) Machine learning-based automatic control of tunneling posture of shield machine. J Rock Mech Geotech Eng 14(4):1153–1164
85. Saadallah A et al (2019) Active learning for accurate settlement prediction using numerical simulations in mechanized tunneling. Procedia CIRP 81:1052–1058
86. Fu H-Y et al (2022) A novel intelligent displacement prediction model of karst tunnels. Sci Rep 12:16983

Chapter 5
A Machine Learning Approach to California Bearing Ratio Prediction: Evaluation of GEP and Ridge Regression

Armin Kashefi Heidarzadeh, S. Danial Mohammadzadeh, Nader Karballaeezadeh, and Reza Yegane Khaksar

Abstract One of the most reliable on-site tests in geotechnical science is the CBR test, whose results play an effective role in estimating the bearing capacity of pavements and in designing them more precisely. The goal of this research is to predict the CBR test using the natural characteristics of the soil. To achieve this aim, Ridge Regression and gene expression programming (GEP) methods are employed to develop a new model that is a function of five independent variables: fine-grained, sand percent, gravel percent, optimum moisture content (OMC), and maximum dry density. The correlation coefficient of the best model resulted in 0.90 and 0.65 for GEP, and 0.90 and 0.70 for Ridge for training and validation data, respectively. The obtained results indicate the high accuracy of the presented model, which can be used for predesign purposes.

Keywords CBR test · Ridge regression · GEP · Predesign · Natural characteristic

---

A. K. Heidarzadeh (✉) · R. Y. Khaksar
Department of Civil Engineering, Sadjad University of Technology, Mashhad, Iran
e-mail: Armin.kashefi121@sadjad.ac.ir

S. Danial Mohammadzadeh
Department of Civil Engineering, Sadjad University of Technology, Mashhad, Iran

Department of Civil Engineering, Faculty of Engineering, Eqbal Lahoori Institute of Higher Education, Mashhad, Iran

N. Karballaeezadeh
Faculty of Civil Engineering, Shahrood University of Technology, Shahrood, Iran

M. Mirrashid et al. (eds.), *Intelligent Predictive Systems*, Emerging Trends in Mechatronics, https://doi.org/10.1007/978-981-95-7766-8_5

5.1 Introduction

Reliable pavement design requires accurate estimation of geomaterial strength indices, among which the California Bearing Ratio (CBR) is commonly used to evaluate subgrade performance. The CBR test, conducted either in the laboratory or in situ, measures the penetration resistance of a soil sample under standardized loading conditions [1–3]. It is defined as the ratio of the applied stress on the soil to the standard stress needed to achieve equivalent penetration in a reference crushed stone material, typically at 2.5 mm or 5 mm penetration depths. The CBR value is calculated as follows:

$$\mathrm{CBR} = \frac{\text{Applied Stress}}{\text{Standard Stress}} * 100 \tag{5.1}$$

This index was first introduced in California in the 1920s and later adopted by the US Army Corps of Engineers in the 1940s for military airfield design [4]. Today, CBR values are widely used in pavement design for highways, airports, and other infrastructures. Furthermore, CBR has been empirically correlated with resilient modulus and other engineering soil properties [5].

Despite its practical value, obtaining reliable CBR measurements poses significant challenges. Financial limitations, inadequate planning, and poor testing conditions can lead to insufficient data quality and inaccurate results [6–8]. The laboratory test itself is laborious and resource-intensive, requiring large quantities of soil—approximately 6 kg per test—and considerable time, especially when numerous tests are needed. These limitations highlight the necessity for alternative methods that can provide reliable estimates with greater efficiency [9].

To address these concerns, researchers have explored predictive modeling approaches using both statistical methods, such as simple and multiple linear regression, and soft computing techniques. Comparative results consistently show that soft computing techniques outperform traditional regression models in predictive accuracy (Table 5.1).

Table 5.1 brings together past studies that have used different computer-based methods to predict CBR values for various soils. Each entry lists the soil type, the number of samples, the modeling approach, and the accuracy of the results (R^2 or R). While the table gives a useful overview of how different methods perform, it

Table 5.1 Details of the previously attempted study for predicting the CBR value

Literature study	Type of soil	Number of datasets	Computational approach	R^2 or R
Thapa and Ghani [10]	Mixed soil samples	100	ANN-GBO	0.95
Khasawneh et al. [11]	Mixed soil samples	110	GA-ELM	0.95
Alawi and Rajab [12]	Granular soil	19	MLR	0.95
Tenpe and Patel [13]	Mixed soil samples	389	GEP ANN	0.82 0.89

also highlights some gaps. Most of the research so far has looked at sands, granular soils, fine-grained soils, or mixed types, while other important soils—like expansive clays or organic soils—are rarely studied. In addition, many of the models rely on more traditional techniques such as multiple linear regression (MLR), artificial neural networks (ANN), and multi-expression programming (MEP). It is obvious that machine learning models have been extensively improving in terms of quality in the areas of geotechnical engineering [14–20]. Therefore, there is room to explore more advanced machine learning methods, and even to combine them with optimization techniques, which could lead to better predictions. Another common issue is that many studies use small datasets, making it harder to apply the findings broadly. Few studies also consider how soil properties can change with location or over time, even though these factors can greatly affect CBR. Lastly, more work is needed to test predictions against real field data and studies that compare different methods across soil types could provide especially valuable insights.

If we look a bit closer, we can see that the number of samples really matters. Some studies only have a handful of data points, which makes it tricky to trust the results completely, while others have hundreds, giving a more solid picture. Also, where the soil comes from is often overlooked. Local conditions like rainfall, temperature, and seasonal changes can affect soil behavior a lot, but many studies don't mention these details. Another thing to notice is that while methods like ANN or GA-ELM can give impressive accuracy, they often don't tell us much about why the soil behaves a certain way. That's why combining these methods with approaches that are easier to interpret could help us understand the soil better while still making good predictions.

On top of that, soil doesn't stay the same over time. Things like compaction, erosion, or changes in the water table can make the soil stronger or weaker, but most models don't take these changes into account. Tracking soil over time and letting models adapt as conditions change could make predictions much more reliable. There's also an exciting chance to use ensemble models—basically combining different approaches to get a more accurate and trustworthy result. Comparing methods across different soil types and situations could really move the field forward and give engineers better tools for real-world work.

Bridging these gaps in the literature can significantly enhance CBR prediction methods, making them more dependable and adaptable across a broader spectrum of soil types and conditions. This, in turn, would directly support advancements in geotechnical engineering and infrastructure development. In this study, Gene Expression Programming (GEP) and Ridge Regression were applied to estimate CBR values. Five soil parameters were selected as input variables: percentage of fines, sand, and gravel, along with maximum dry density and optimum moisture content. The accuracy of the developed models was assessed using a combination of statistical measures and graphical evaluations, while additional statistical indices were used to examine their generalization capability. Nonetheless, some limitations should be acknowledged. Since the dataset was obtained from local transportation authorities, potential regional bias may affect the broader applicability of the results.

Moreover, natural variability in soil properties could also influence the reliability of the predictions.

5.2 Materials and Methodology

5.2.1 Data Gathering

The dataset, obtained from the Soil Mechanics and Technical Laboratory of Razavi Khorasan Province, Iran, includes 101 samples, with 72 samples (70%) used for training the model and 29 samples (30%) for validation. This division ensured balanced exposure to data for model development and unbiased evaluation during testing [21, 22]. The study focuses on critical geotechnical features that directly impact soil strength behavior, including fine content (*F*), sand percentage (*S*), gravel percentage (*G*), maximum dry density (MDD), and optimum moisture content (OWC). These variables reflect the soil's grading and compaction characteristics, which are essential for predicting its bearing capacity. Statistical summaries for these parameters across training and validation sets are provided in Tables 5.2 and 5.3, while Fig. 5.1 illustrates their distributions.

In addition to providing a foundation for modeling, this dataset captures the natural variability of soils from the Razavi Khorasan region, reflecting differences in texture, composition, and compaction characteristics. Such variability is important

Table 5.2 Statistical characteristics of the training dataset

		Indicators			
Variables	Category	Min	Max	Avg	St. dev.
F	Input	2.6	30.3	11.81	6.07
S	Input	24.2	63.3	46.25	9.92
G	Input	15.9	72.3	41.93	13.21
OWC	Input	4	9.5	6.11	1.05
MDD	Input	2.08	2.9	2.27	9.36
CBR	Output	49	116	83.25	14.92

Table 5.3 Statistical characteristics of the validation dataset

		Indicators			
Variables	Category	Min	Max	Avg	St. dev.
F	Input	3.3	27.2	13.52	5.8
S	Input	19.1	63.3	46.25	9.92
G	Input	20	48.3	44.4	11.86
OWC	Input	4.2	8.1	6.02	0.86
MDD	Input	2.185	2.41	2.28	5.04
CBR	Output	67	102	84.38	10.54

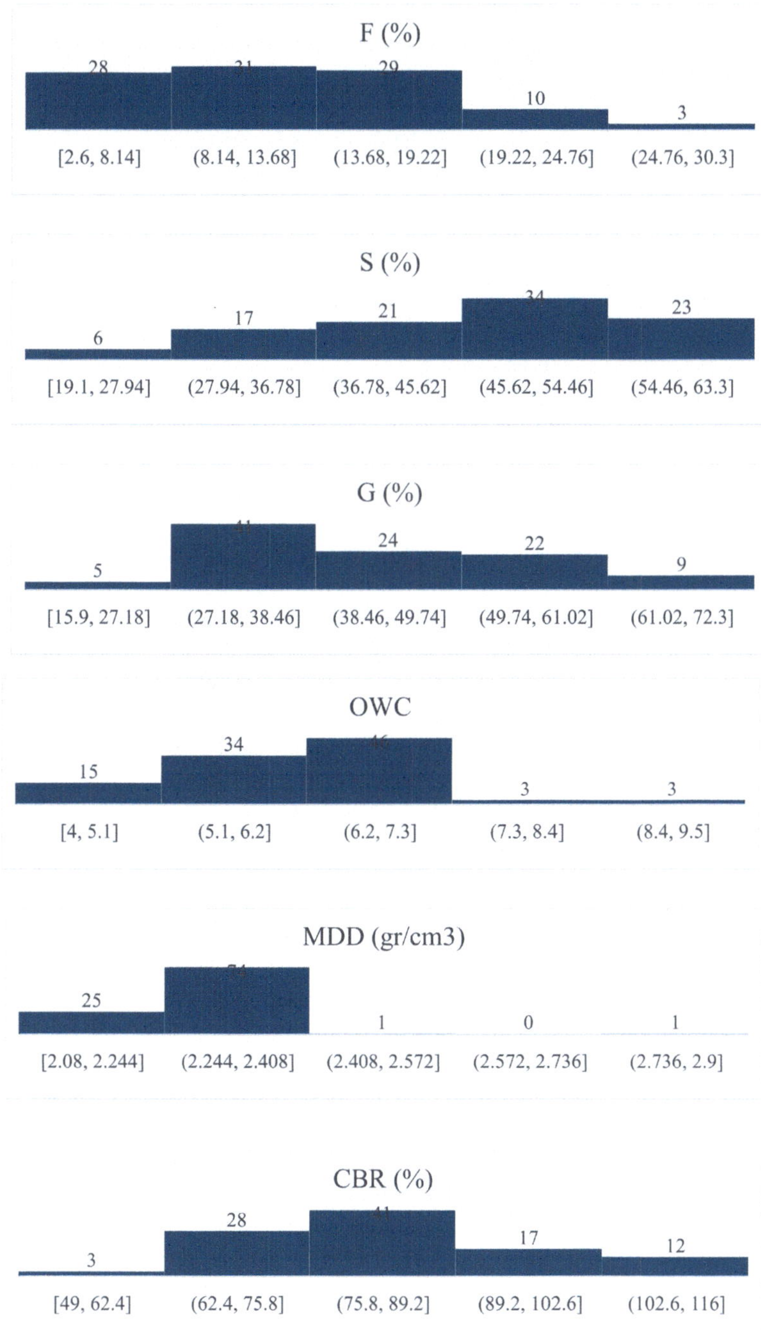

Fig. 5.1 Marginal distribution histograms of the variables existing in the dataset

for training models that can generalize to new, unseen soil samples. By examining the distributions of these parameters, researchers can identify trends, potential outliers, and relationships between variables, which can offer insights into how soil composition influences bearing capacity.

Moreover, analyzing the dataset in detail allows for better understanding of how individual factors contribute to soil strength. For example, the interaction between fine content and moisture levels can significantly affect compaction and load-bearing capacity, while the proportions of sand and gravel influence granular stability. These insights are not only valuable for model development but also provide practical information for field engineers assessing soil suitability for construction.

Finally, the division into training and validation sets ensures that the model's predictive performance can be evaluated rigorously. By reserving a portion of data for validation, it becomes possible to assess the model's accuracy on unseen samples, detect potential overfitting, and refine the model's parameters to improve reliability. This careful approach to dataset handling lays the groundwork for robust predictive modeling and provides confidence in the applicability of the results to real-world geotechnical engineering problems.

GEP and Ridge Regression were applied to capture complex relationships among variables. The GEP model was tuned through a trial-and-error approach, while Ridge Regression employed cross-validation for hyperparameter optimization. GEP was chosen for its ability to generate explicit mathematical expressions describing nonlinear soil behavior, whereas Ridge Regression was selected for its robustness against multicollinearity and overfitting.

In Ridge Regression, input variables were standardized using the StandardScaler function in scikit-learn to ensure equal weighting of features. A dataset of 101 observations representing varied soil conditions was used for training and validation. The models were implemented with GeneXproTools 50 for GEP, taking advantage of its symbolic regression and genetic programming capabilities, and Python in the Google Colab environment for Ridge Regression, using libraries such as scikit-learn for training, tuning, and validation.

Beyond these technical details, applying both GEP and Ridge Regression allows us to explore soil behavior from two complementary angles. GEP provides formulas that can be interpreted and applied directly, giving a sense of "why" the soil behaves in a certain way, while Ridge Regression offers stability and reliability, even when some variables are closely correlated. Using these two approaches together gives a richer, more nuanced understanding of the dataset.

The trial-and-error tuning of GEP, although time-consuming, was essential for identifying the most effective model structure. Each adjustment helped capture subtle nonlinear patterns in the soil properties that simpler methods might miss. Meanwhile, Ridge Regression's cross-validation approach ensured that the model didn't just fit the training data well, but could also make reliable predictions on unseen samples.

Implementing the models in GeneXproTools and Python also brought practical advantages. GeneXproTools' symbolic regression makes it easy to visualize relationships in a way that's intuitive and actionable for engineers, while Python's ecosystem provides flexibility and powerful libraries for handling data preprocessing, model training, and validation. Together, these tools create a workflow that is both rigorous and adaptable, reflecting real-world engineering practices where datasets vary and models need to be interpretable and dependable.

5.2.2 Genetic Expression Programming (GEP)

Gene Expression Programming (GEP) is a modern variant of Genetic Programming (GP) that provides a flexible and powerful framework for modeling complex systems through symbolic regression. While traditional GP evolves solutions in a tree-like structure inspired by biological evolution, GEP approaches the problem differently. It represents candidate solutions using fixed-length linear strings, known as chromosomes, which are then translated into expression trees (ETs) of varying sizes and shapes, allowing the evolutionary process to maintain both structural adaptability and genetic diversity throughout generations. This characteristic enables GEP to capture nonlinear interactions among input variables while preserving interpretability [23–32].

First introduced by Ferreira in 2002, GEP is structured around essential components including terminals, functions, control parameters, a fitness function, and a termination condition. Unlike conventional GP, which operates directly on tree structures, GEP encodes its models in multiple genes within a chromosome, where each gene functions as a self-contained subprogram. These subprograms evolve in parallel and collectively form the final solution, allowing for modular and hierarchical modeling of complex relationships [23–32].

Each gene is divided into two parts: the head, which may contain both function and terminal symbols, and the tail, which contains only terminals. Together, they determine the behavior and structure of the expression tree. GEP supports a wide variety of operations, from basic arithmetic to logarithmic, exponential, and inverse functions, which enhances its capacity to model nonlinear dependencies in geotechnical and engineering problems. The linear portion of the resulting models, including coefficients such as c_0, c_1, and c_2, can be calibrated using least squares methods. Depending on the modeling context and dataset characteristics, alternative calibration techniques, such as partial least squares, can also be employed to refine the model accuracy [23–32].

One of the key advantages of GEP is its ability to produce concise, closed-form solutions. Careful selection of parameters, such as tree depth and the number of genes, allows for control over model complexity, balancing accuracy and interpretability. Shallower trees typically result in shorter, more compact equations, facilitating easier implementation in real-world engineering applications. This feature makes GEP models both powerful and practical, providing engineers and researchers with transparent mathematical representations of the relationships among soil properties and other input variables [23–32].

Furthermore, the flexibility of GEP in encoding multiple genes and the possibility of using different function sets enable the model to explore a wide solution space, reducing the risk of premature convergence and improving generalization on unseen data [23, 24, 33]. The ability to generate interpretable expression trees while maintaining predictive performance distinguishes GEP from conventional GP, making it particularly suitable for engineering applications where both understanding and accuracy are critical [23–32].

5.2.3 *Ridge Regression*

Ridge regression, first introduced by Hoerl and Kennard in 1970, was developed to make regression models more stable and reliable when predictor variables are strongly correlated. In such cases, ordinary least squares (OLS) often produce unstable estimates, which can lead to poor predictions. Ridge regression addresses this by adding an L2 penalty to the model—this penalty gently pulls the coefficients towards zero, reducing variance while slightly increasing bias to achieve better overall performance [34, 35]. Mathematically, the ridge estimator is expressed as:

$$\hat{\beta}_{\text{ridge}} = \left(X^{\mathrm{T}} X + \lambda I\right)^{-1} X^{\mathrm{T}} y \tag{5.2}$$

Ridge regression relies on a non-negative tuning parameter, λ, which controls the degree of shrinkage applied to the coefficients. When $\lambda = 0$, the method reduces to OLS, but as λ increases, the coefficients are drawn closer to zero, producing more stable results, particularly in cases of multicollinearity [36, 37]. This shrinkage stabilizes the inversion of $X^{\mathrm{T}}X$ and mitigates inflated variances that otherwise distort coefficient estimates, making ridge regression especially powerful for ill-conditioned or high-dimensional datasets [35, 37].

Over the years, the ridge framework has evolved through adaptive and data-driven strategies. López et al. (2024) introduced an adaptive ridge method that modifies the penalty based on data structure, improving predictive accuracy [34]. Similarly, He (2024) extended ridge regression to dense factor-augmented models, confirming its robustness and consistency in complex data environments [38]. Sparse Ridge Regression further builds on this principle by incorporating an additional sparsity-inducing term, allowing certain coefficients to shrink entirely to zero and improving interpretability without losing ridge stability [39].

Applications of ridge regression span a wide range of fields. In evolutionary biology, Sztepanacz (2024) applied it to estimate multivariate selection patterns while preventing overfitting in high-dimensional trait datasets [40]. In engineering, Li (2024) combined ridge regression with piecewise fitting to analyze oil and gas reservoir data, effectively capturing nonlinear behavior amid measurement noise [33]. The same regularization principle has been implemented in machine learning frameworks such as scikit-learn [41], ensuring that models remain generalizable rather than overfitted [35].

Taken together, these developments highlight ridge regression's value as a balanced and enduring approach—one that elegantly trades off bias for reduced variance and continues to inspire new hybrid techniques for stable and interpretable modeling.

5.2.4 Performance Evaluation Methods

In light of their correlations and error rates, this study presents several evaluation criteria for employed models. Correlation (R^2), mean absolute error (MAE), and root mean squared error (RMSE) are among the evaluation metrics examined. The formulas that apply to each of these metrics are listed below [31, 32]. In both training and validation phases, an algorithm performs exceptionally well if it attains a high R^2 value close to 1. Metrics that indicate that the model has less error, such as RMSE and MAE, have smaller values and are, therefore, preferred.

These metrics together offer a more complete picture of model performance. While a high R^2 demonstrates that the model captures most of the variability in the target variable, MAE and RMSE provide insight into the magnitude of prediction errors, ensuring that outliers or extreme deviations are taken into account. Evaluating both correlation and error metrics allows for a balanced assessment, highlighting not only overall fit but also precision in individual predictions.

Assessing multiple metrics also facilitates comparison between different modeling approaches. For instance, a model might achieve a slightly higher R^2 but display larger RMSE values, indicating fewer stable predictions for certain samples. Conversely, a model with marginally lower correlation but smaller errors may offer more reliable results in practice. Considering all these factors together helps identify the most suitable model for real-world geotechnical applications, where both accuracy and robustness are critical.

$$R^2 = \left(\frac{\sum_{i=1}^{M} (p_i - \bar{p})(I_i - \bar{I})}{\sqrt{\left[\sum_{i=1}^{M} (p_i - p)^2\right]\left[\sum_{i=1}^{M} (I_i - \bar{I})^2\right]}} \right)^2 \tag{5.3}$$

$$\text{RMSE} = \sqrt{\frac{1}{M} \sum_{i=1}^{M} (I_i - p_i)^2} \tag{5.4}$$

$$\text{MAE} = \frac{1}{M} \sum_{i=1}^{M} |I_i - p_i| \tag{5.5}$$

Equations (5.3)–(5.5) employ the notation M to denote the total number of samples, p_i to represent the predicted value, and l to signify the mean of both predicted and measured values, respectively. Meanwhile, l_i refers to the measured value.

5.3 Results and Discussion

In this section, the predictive performance of the two developed models is presented and discussed in detail. In addition, graphical comparisons between the predicted outcomes and the laboratory-measured values are provided for both the training and validation datasets. The only hyperparameter considered for Ridge Regression is alpha. This hyperparameter focuses on controlling the degree of shrinkage applied to the regression coefficients, and its task is to stabilize the solution by reducing the effect of multicollinearity and preventing overfitting. Table 5.4 provides three candidate values for alpha (0.1, 1, 10). Finally, alpha = 0.1 was chosen.

In the hyperparameter tuning process, several candidate values for the Ridge regression regularization parameter α were considered: 0.1, 1, and 10. For each value, the model was trained on the training dataset and evaluated on the validation dataset using R^2, RMSE, and MAE. The α value that provided the best predictive performance (highest R^2 and lowest RMSE) on the validation set was selected as the optimal hyperparameter. All final results reported in this study correspond to the Ridge regression model tuned with this optimal α, which is 0.1.

The GEP algorithm was run multiple times using different combinations ($3 \times 3 \times 3 \times 3 \times 3 \times 2 = 486$) of parameters to find the most effective model. Parameter ranges were guided by previous studies [34–36], and Table 5.5 shows the specific settings used. Larger population sizes and numbers of generations were chosen to give the algorithm the best chance of finding optimal solutions. Each individual's fitness was evaluated using roulette wheel selection with elitism, while genetic operators like crossover, mutation, and rotation helped maintain diversity in the population. Since the best parameter settings, such as the number of genes and chromosomes, gene head size, and operator rates, can greatly affect performance, they were fine-tuned through trial and error.

The extensive exploration of parameter combinations ensured that the GEP model could capture the complex, nonlinear relationships among soil properties effectively. Trial-and-error tuning allowed for careful adjustment of key parameters, improving both the accuracy and stability of predictions. By using larger populations and more generations, the algorithm had more opportunities to evolve robust solutions, while the application of genetic operators maintained variability, preventing premature convergence on suboptimal results.

Table 5.4 Parameters employed in the development of the Ridge Regression model

Parameters	Setting
Alpha (α or λ)	0.1, 1, 10

Table 5.5 Parameters employed in the development of the GEP-based model and the best model

Parameters	Setting	Tuned Values
Function sets	+, −, *, /, POWER, Exp, Sqrt, Ln, Sin, Cos, Tan	+, −, *, /, POWER, Exp, Sqrt, Ln, Sin, Cos, Tan
Size of population	100, 150, 200	200
Number of generations	1000, 1500, 2000	2000
Chromosome length	50, 100, 200	200
Number of genes	3–5	5
Head size	6, 8, 10	10
Linking function	+	+
One-point recombination rate (%)	20, 30	30
Two-point recombination rate (%)	30	30
Gene recombination rate (%)	7	7
Gene transposition rate (%)	70	70
Numerical constants	Floating	Floating

Table 5.5 not only summarizes the parameter settings but also highlights the best values determined during tuning. These optimized settings provide a balance between computational efficiency and model performance, ensuring that the final GEP-based model is both reliable and interpretable for practical geotechnical applications.

Based on the provided parameters, the performance of the Ridge Regression model in predicting the California Bearing Ratio value of granular soil under dry conditions is illustrated in Figs. 5.2 and 5.3. Figure 5.2 compares the actual CBR values from laboratory tests with the predicted values from the Ridge Regression model for the validation dataset, highlighting the predictive performance of the tuned model.

The trace of coefficients plot (Fig. 5.3) illustrates how the magnitude of each feature's coefficient changes with increasing α (λ) in the Ridge regression model. As α increases, the coefficients shrink towards zero, demonstrating the effect of regularization: higher α values penalize less important coefficients to prevent overfitting. In the plot, features such as *F*, *S*, and *G* maintain relatively high coefficient values even for larger α, indicating that they are the most influential features for predicting the target variable. Conversely, features like OWC and MDD decrease rapidly towards zero as α grows, showing that they are less influential in the model. Therefore, the trace of coefficients plot not only highlights the sensitivity of each feature to regularization but also clearly distinguishes the most important features from the less significant ones, providing insight into their relative contribution to the prediction.

By observing how coefficients respond to varying α, it becomes clear which soil properties have the strongest influence on the CBR value. Features such as *F*, *S*, and *G* retain relatively high coefficients even at larger α values, emphasizing their dominant role in predicting the target variable. In contrast, OWC and MDD coefficients decrease quickly, showing that these features contribute less to the model's predictions.

The trace of coefficients plot thus serves as a valuable diagnostic tool, not only highlighting the effect of regularization but also providing insight into the relative importance of each feature. Understanding which variables are most influential can guide engineers and researchers in focusing on critical soil properties, optimizing data collection efforts, and improving future predictive models.

Overall, combining visual assessments like Fig. 5.2 with coefficient analysis in Fig. 5.3 offers a comprehensive understanding of model performance, feature significance, and the practical implications of Ridge Regression in geotechnical engineering applications.

The results of the GEP model are presented in Figs. 5.4 and 5.5. Figure 5.4 compares the actual and predicted values of the validation dataset, *highlighting the model's predictive performance*, while Fig. 5.5 depicts the expression tree generated by the model.

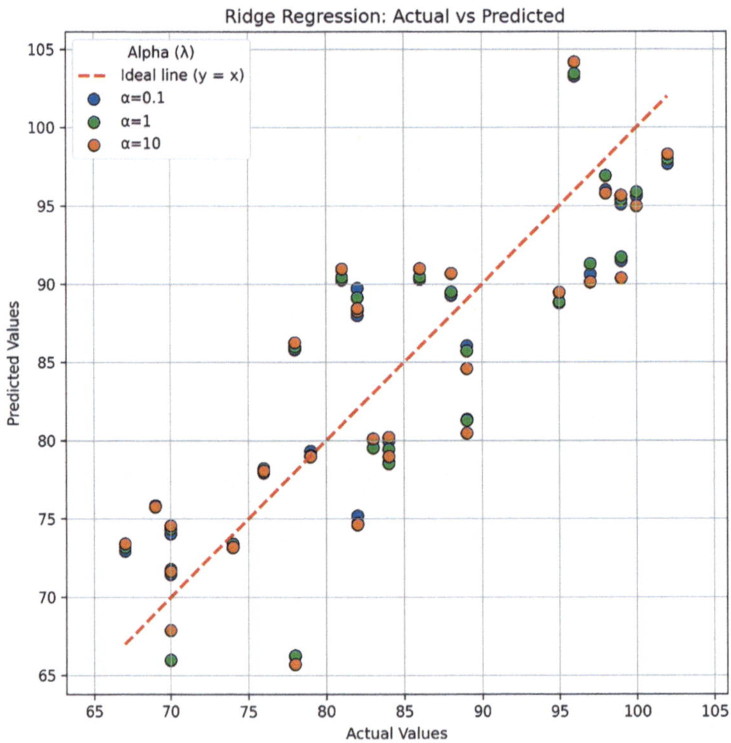

Fig. 5.2 Actual versus predicted CBR values for the validation subset using the ridge regression model tuned with optimal hyperparameters

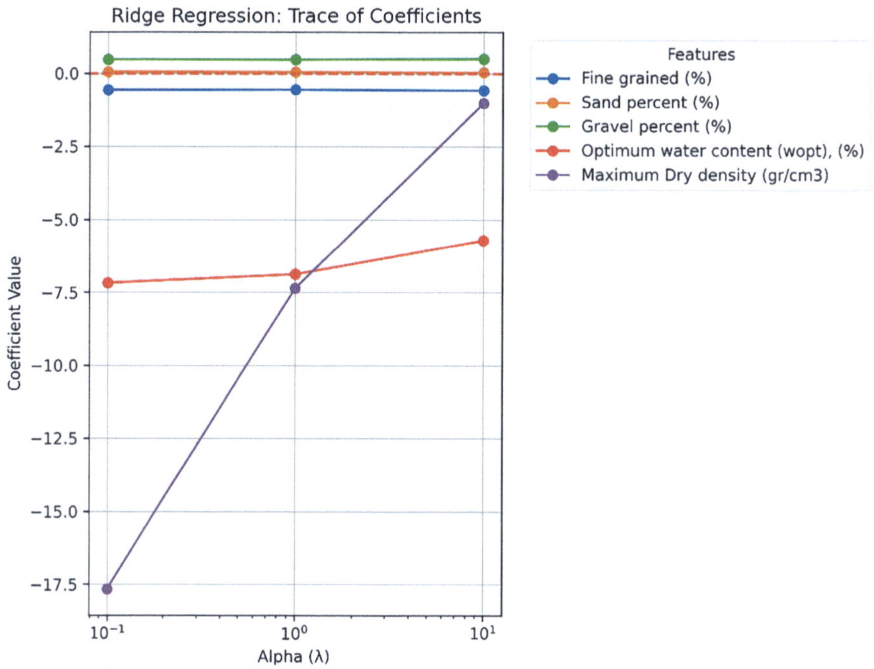

Fig. 5.3 Ridge regression coefficients for the validation subset plotted against different α values

Figure 5.4 further demonstrates the model's effectiveness by comparing predicted and actual CBR values. The close alignment between the two sets of values indicates that the GEP model successfully captures the complex, nonlinear interactions among soil parameters. Deviations from the ideal 1:1 line are minimal, suggesting high accuracy and robustness of the model when applied to unseen validation data.

The expression tree in Fig. 5.5 provides a clear visualization of the mathematical relationships captured by the GEP model. Each branch represents a function or operation, while the leaves correspond to the input variables, offering a transparent view of how soil properties are combined to predict the CBR value. This interpretability is one of the key advantages of GEP, allowing engineers to understand not just the output but also the underlying decision process.

By examining Figs. 5.4 and 5.5 together, it becomes clear that the GEP model combines both interpretability and predictive power. The expression tree allows researchers to trace exactly how input variables influence predictions, while the visual comparison in Fig. 5.4 confirms the reliability of the model. Such insights can guide engineers in identifying the most critical soil properties affecting bearing capacity.

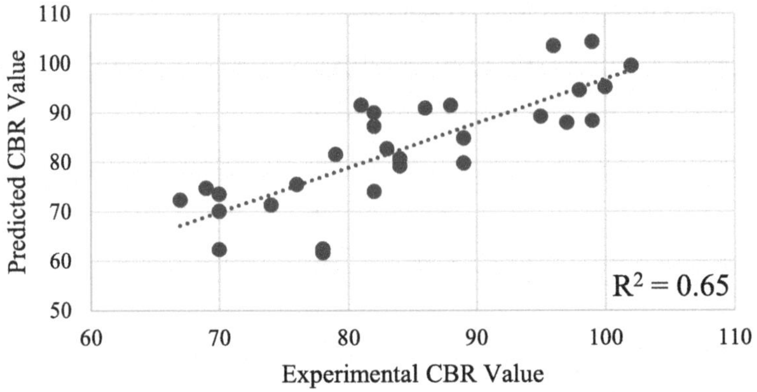

Fig. 5.4 Comparison of predicted and experimental CBR values for the validation subset

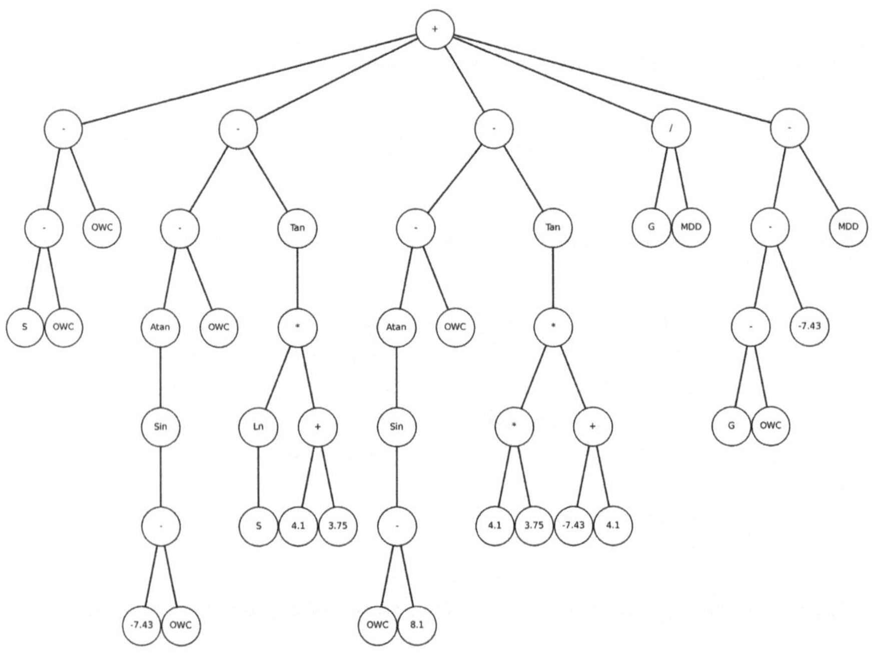

Fig. 5.5 Illustration of the GEP program, representing the model inputs and corresponding coefficients

These visualizations also serve an educational purpose, making it easier to communicate complex model behavior to a broader audience. They highlight how advanced computational models can be both accurate and interpretable, which is particularly valuable for practical decision-making in geotechnical engineering projects.

Additionally, the following closed-form equation was derived and utilized to develop the GEP-based model:

$$\begin{aligned}
\mathrm{CBR} = {} & (S - 3 * \mathrm{OWC}) + \begin{pmatrix} \arctan\left(\sin\left(-7.43 - \mathrm{OWC}\right)\right) \\ -\mathrm{OWC} - \tan\left(\ln(S) * 7.85\right) \end{pmatrix} \\
& + \left(\arctan\left(\sin\left(\mathrm{OWC} - 8.1\right)\right) - \mathrm{OWC} - \tan\left(-26.14\right)\right) \\
& + (G - \mathrm{MDD}) + (G - \mathrm{OWC} + 7.43 - \mathrm{MDD})
\end{aligned} \tag{5.6}$$

Although F was included among the inputs, it does not appear in the final GEP equation. This indicates that F contributed little independent information to the prediction of CBR, likely because its effect was already captured by other correlated soil parameters.

The predictive capability of the developed models for estimating CBR was assessed using key performance metrics—RMSE, R^2, and MAE—across both training and validation datasets. The comparison is illustrated, with detailed numerical results summarized in Tables 5.6 and 5.7.

Table 5.6 presents the results of the developed GEP model, while Table 5.7 summarizes the performance of the Ridge Regression model. For the GEP model, the training phase shows strong predictive capability, with an RMSE of 4.70, an R^2 of 0.901, and an MAE of 3.76. During validation, the model's performance declines slightly, with RMSE rising to 7.06, R^2 dropping to 0.65, and MAE increasing to 5.90. This decrease is likely due to mild overfitting, as the GEP model may have captured some dataset-specific patterns that do not fully generalize to unseen samples. Additionally, the validation dataset may include samples with higher measurement noise or values outside the range observed during training.

Ridge Regression, however, exhibits remarkably consistent performance across all tested α values. In the validation phase, it achieves the lowest RMSE, highest R^2, and lowest MAE compared with GEP, indicating superior generalization capability and resilience against overfitting. For example, with $\alpha = 0.1$, Ridge Regression attains an RMSE of 5.68, R^2 of 0.70, and MAE of 5.01, showing that even modest regularization produces reliable predictions. Increasing α slightly affects error metrics, but the model maintains stable performance, highlighting its robustness.

Overall, the comparison between GEP and Ridge Regression underscores important considerations for model selection in geotechnical engineering. While GEP provides interpretable mathematical expressions and captures nonlinear relationships effectively, Ridge Regression offers slightly higher predictive accuracy and

Table 5.6 Results of the developed GEP model

		Index values		
Model	Phase	RMSE	R^2	MAE
GEP	Training	4.70	0.901	3.76
	Validation	7.06	0.65	5.90

Table 5.7 Results of the developed Ridge Regression model

			Index values		
Model	Phase	α	RMSE	R^2	MAE
Ridge regression	Training	0.1	4.59	0.90	3.91
		1.0	4.74	0.90	4.02
		10	5.00	0.89	4.16
Ridge regression	Validation	0.1	5.68	0.70	5.01
		1.0	5.71	0.70	5.03
		10	5.93	0.67	5.20

more stable performance across different conditions. This illustrates a trade-off between interpretability and predictive reliability that engineers must consider when choosing a computational model for practical applications.

The results also demonstrate the value of using multiple evaluation metrics, including RMSE, R^2, and MAE, to obtain a comprehensive view of model performance. Each metric provides a different perspective: RMSE emphasizes larger errors, MAE provides average error magnitude, and R^2 indicates overall correlation. Considering all three together ensures that the selected model is both accurate and reliable for predicting the California Bearing Ratio under varied soil conditions.

These findings provide actionable insights for geotechnical engineers, suggesting that Ridge Regression may be preferable when consistent predictive performance is critical, while GEP remains valuable when interpretability and understanding of the underlying relationships are priorities. Such insights can guide both the development of predictive tools and their practical deployment in field assessments of soil bearing capacity.

5.4 Conclusion

This study proposes an advanced framework for predicting the dry California Bearing Ratio (CBR) of granular soils by utilizing two machine learning techniques: Gene Expression Programming and Ridge Regression. Soil samples were collected from several ongoing highway construction projects in Mashhad, Razavi Khorasan, Iran, and their geotechnical properties were determined through standardized laboratory procedures.

The final models incorporated five key input variables, reflecting the primary soil characteristics influencing CBR. Both GEP and Ridge Regression achieved satisfactory predictive capability, with strong correlation coefficients for the training dataset. The GEP model demonstrated low prediction errors, and Ridge Regression showed slightly improved performance, confirming the reliability of both approaches. Validation using unseen data indicated consistent generalization, with Ridge Regression performing marginally better than GEP.

The findings of this research highlight the practical value of integrating evolutionary algorithms with statistical regression techniques for predicting CBR. The models offer a faster and more cost-effective alternative to conventional laboratory testing while capturing the essential relationships among soil properties.

Future research could focus on applying these models to a wider variety of soil types, exploring additional optimization methods, or integrating them with real-time monitoring systems to enhance adaptability and predictive reliability. Overall, this study demonstrates that advanced computational models can effectively support geotechnical practice, providing reliable predictions of soil bearing capacity for engineering applications.

AI Use Disclosure Statement Artificial intelligence tools were used exclusively for proofreading and improving language clarity. Specifically, ChatGPT (OpenAI) was employed to correct grammar and enhance readability. All ideas, analyses, and conclusions presented in this chapter are entirely the author's own.

References

1. Haupt F, Netterberg F (2021) Prediction of California bearing ratio and compaction characteristics of Transvaal soils from indicator properties. J South Afr Inst Civ Eng 63(2):47–56
2. Katte VY et al (2019) Correlation of California bearing ratio (CBR) value with soil properties of road subgrade soil. Geotech Geol Eng 37(1):217–234
3. Nagaraju TV, Prasad CD, Raju MJ (2019) Prediction of California bearing ratio using particle swarm optimization. In: Soft computing for problem solving: SocProS 2018, vol 1. Springer, pp 795–803
4. Purwana Y, Nikraz H, Jitsangiam P (2012) Experimental study of suction-monitored CBR test on sand-kaolin clay mixture. GEOMATE J 3(6):419–422
5. Ahmad M et al (2023) Predicting California bearing ratio of HARHA-treated expansive soils using Gaussian process regression. Sci Rep 13(1):13593
6. Black W (1962) A method of estimating the California bearing ratio of cohesive soils from plasticity data. Geotechnique 12(4):271–282
7. Putra H, Yudhistira I (2022) Improvement of the California bearing ratio of peat soil using soybean crude urease calcite precipitation. Civil Eng J 8(11):2411–2423
8. Mohammadzadeh DS et al (2025) Analysis and modeling of soil retaining structure for urban train stations. In: Advancements in underground infrastructures. CRC Press, pp 147–183
9. Verma G, Kumar B (2022) Application of multi-expression programming (MEP) in predicting the soaked California bearing ratio (CBR) value of fine-grained soil. Innov Infrastruct Solut 7(4):264
10. Thapa I, Ghani S (2024) Estimation of California bearing ratio for hill highways using advanced hybrid artificial neural network algorithms. Multiscale Multidiscip Model Exp Des 7(2):1119–1144
11. Khasawneh MA et al (2024) Integrating soil index parameters and machine learning for reliable California bearing ratio estimation using a GA-ELM model. Int J Pavement Res Technol:1–18. https://doi.org/10.1007/s42947-024-00489-x
12. Alawi M, Rajab M (2013) Prediction of California bearing ratio of subbase layer using multiple linear regression models. Road Mater Pavement Des 14(1):211–219
13. Tenpe AR, Patel A (2020) Application of genetic expression programming and artificial neural network for prediction of CBR. Road Mater Pavement Des 21(5):1183–1200

14. Maroof A, Karballaeezadeh N, Mohammadzadeh SD, Armaghani DJ, Mirzaei H (2025) Prediction of the wetting-induced compression of collapsible soils using ensemble machine learning. Indian Geotechn J:1–20
15. Jalali H, Yeganeh Khaksar R, Mohammadzadeh SD, Karballaeezadeh N, Gandomi AH (2024) Prediction of vertical displacement for a buried pipeline subjected to normal fault using a hybrid FEM-ANN approach. Front Struct Civ Eng 18(3):428–443
16. Armaghani DJ, Mohamad ET, Narayanasamy MS, Narita N, Yagiz S (2017) Development of hybrid intelligent models for predicting TBM penetration rate in hard rock condition. Tunn Undergr Space Technol 63:29–43
17. Afrazi M, Jahed Armaghani D, Afrazi H, Fattahi H, Samui P (2025) Real-time monitoring of tunnel structures using digital twin and artificial intelligence: a short overview. Deep Underground Sci Eng.:1–16. https://doi.org/10.1002/dug2.70029
18. He B, Armaghani DJ, Lai SH, He X, Asteris PG, Sheng D (2024) A deep dive into tunnel blasting studies between 2000 and 2023—a systematic review. Tunn Undergr Space Technol 147:105727
19. Asteris PG, Armaghani DJ (2025) An empirical-driven machine learning (EDML) approach to predict PPV caused by quarry blasting. Bull Eng Geol Environ 84(4):200
20. Murlidhar BR, Nguyen H, Rostami J, Bui XN, Armaghani DJ, Ragam P, Mohamad ET (2021) Prediction of flyrock distance induced by mine blasting using a novel Harris Hawks optimization-based multi-layer perceptron neural network. J Rock Mech Geotech Eng 13:1413–1427
21. Qian Y et al (2024) Prediction of ultra-high-performance concrete (UHPC) properties using gene expression programming (GEP). Buildings 14(9):2675
22. Maroof A et al (2025) Prediction of the wetting-induced compression of collapsible soils using ensemble machine learning. Indian Geotech J:1–20
23. Ferreira C (2001) Gene expression programming: a new adaptive algorithm for solving problems. Complex Syst 13:87–129
24. Zhou C, Xiao W, Tirpak TM, Nelson PC (2002) Discovery of classification rules by using gene expression programming. In: Proceedings of the 2002 international conference on artificial intelligence (IC-AI'02), Las Vegas, Jun 2002, pp 1355–1361
25. Mohammadzadeh SD et al (2019) Prediction of compression index of fine-grained soils using a gene expression programming model. Infrastructures 4(2):26
26. Gardezi H et al (2024) Predictive modeling of rutting depth in modified asphalt mixes using gene-expression programming (GEP): a sustainable use of RAP, fly ash, and plastic waste. Constr Build Mater 443:137809
27. Mohammadzadeh D, Bazaz JB, Alavi AH (2014) An evolutionary computational approach for formulation of compression index of fine-grained soils. Eng Appl Artif Intell 33:58–68
28. Khawaja L et al (2024) Indirect estimation of resilient modulus (Mr) of subgrade soil: gene expression programming vs multi expression programming. In: Structures. Elsevier
29. Vargas R, Mosavi A, Ruiz R (2017) Deep learning: a review. Adv Intell Syst Comput 7:122–148
30. Mosavi A, Rabczuk T, Varkonyi-Koczy AR (2017) Reviewing the novel machine learning tools for materials design. In: Proceedings of the international conference on global research and education, Lasi, Romania, 23 Sept 2017, pp 50–58
31. Ferreira C (2002) Gene expression programming in problem solving. In: Soft computing and industry. Springer, Angra do Heroismo, pp 635–653
32. Mosavi A, Rabczuk T (2017) Learning and intelligent optimization for material design innovation. In: Proceedings of the international conference on learning and intelligent optimization, Russia, Novgorod, 19 Jun 2017, pp 358–363
33. Li Y et al (2024) Regression analysis and its application to oil and gas exploration: a case study of hydrocarbon loss recovery and porosity prediction, China. Energy Geosci 5(4):100333
34. Montesinos-López A et al (2024) Refining penalized ridge regression: a novel method for optimizing the regularization parameter in genomic prediction. G3 Genes Genomes Genet 14(12):jkae246

35. Yahya WB, Olaifa J (2014) A note on ridge regression modeling techniques. Electron J Appl Stat Anal 7:343–361. https://doi.org/10.1285/i20705948v7n2p343
36. Hoerl AE, Kennard RW (1970) Ridge regression: biased estimation for nonorthogonal problems. Technometrics 12(1):55–67
37. Alkinani HH, Al-Hameedi ATT, Dunn-Norman S et al (2022) Regularized ridge regression models to estimate static elastic moduli from wireline measurements: case study from Southern Iraq. J Petrol Explor Prod Technol 12:1759–1773. https://doi.org/10.1007/s13202-021-01434-5
38. He Y (2024) Ridge regression under dense factor augmented models. J Am Stat Assoc 119(546):1566–1578
39. Huang W, Cai Y, Zhang G (2025) Battery degradation analysis through sparse ridge regression. Energy Syst. https://doi.org/10.71070/es.v4i1.65
40. Sztepanacz JL, Houle D (2024) Regularized regression can improve estimates of multivariate selection in the face of multicollinearity and limited data. Evol Lett 8(3):361–373
41. Pedregosa F, Varoquaux G, Gramfort A, Michel V, Thirion B, Grisel O et al (2011) Scikit-learn: machine learning in Python. J Mach Learn Res 12:2825–2830

Chapter 6
Data-Driven Modeling of Strength Properties in Waste-Stabilized Soils: A Review of Artificial Intelligence Approaches

Jitendra Khatti, Leyla Bouaricha, and Sidali Denine

Abstract The inclusion of industrial and agricultural by-products into soil stabilization offers a sustainable alternative to traditional binders like cement, which have a large environmental impact. Nonetheless, the strength behavior of soils treated with waste is not easily predictable due to complex nonlinear relations among soil constituent materials and the stabilizing agent(s). This presents a systematic review of the convergence of two research areas: (1) use of different waste products to improve geotechnical performance, and (2) application of artificial intelligence (AI) techniques to model soil behavior. This review synthesizes evidence on the use of materials such as fly ash, rice husk ash, refined granulated blast-furnace slag, and alum sludge, assessing the influence on important mechanical parameters, generally unconfined compressive strength (UCS) and California Bearing Ratio (CBR). Advanced AI and machine learning capabilities, such as Artificial Neural Networks (ANN), Genetic Programming (GP), and Extreme Gradient Boosting (XGBoost). Findings reveal that AI-based frameworks yield high accuracy ($R^2 > 0.95$) and also frequently outperform empirical methods, while offering the strength of data integration. With some limitations concerning data availability, interpretability, and field applicability, AI-driven methods present considerable promise in the improved performance of soil-waste unit soils and more circular economies in geotechnical engineering.

J. Khatti
Department of Civil Engineering, Rajasthan Technical University, Kota, Rajasthan, India

L. Bouaricha (✉)
Department of Civil Engineering and Architecture, Faculty of Sciences and Technology, University of Khemis Miliana, Khemis Miliana, Algeria
e-mail: l.bouaricha@univ-dbkm.dz

S. Denine
Laboratory of Materials Science and Environment (LMSE), Hassiba Benbouali University of Chlef, Chlef, Algeria

Civil Engineering Department, University Center of Tipaza, Tipaza, Algeria
e-mail: s.denine@univ-chlef.dz

M. Mirrashid et al. (eds.), *Intelligent Predictive Systems*, Emerging Trends in Mechatronics, https://doi.org/10.1007/978-981-95-7766-8_6

Keywords Soil stabilization · Waste materials · Artificial intelligence · Machine learning · Sustainable geotechnics

6.1 Introduction

Soil stabilization is a key process in civil engineering for specifying the stability, durability, and load-bearing capacity of structures built on weak or low-quality soils [1]. The process of soil stabilization can be achieved through a range of stabilization methods, where (1) mechanical methods to (2) chemical methods that involve the use of different additives [2]. The traditional stabilizers such as lime and cement have been the most widely used for improving soil properties [3, 4]. These stabilizers work effectively, but have a high environmental cost, which produces approximately 5% of all global CO_2 emissions [5–7]. Therefore, considering the dual challenges of climate change and waste disposal, industrial by-products and agricultural wastes are introduced as sustainable alternatives for soil stabilization and are defined as "green materials" [8].

Materials such as fly ash, rice husk ash (RHA), waste glass powder (WGP), ground-granulated blast-furnace slag (ggbfs), and alum sludge are waste materials (or by-products) that can (1) help mitigate urgent waste disposal issues while (2) technically improving the engineering properties of soils [7, 9]. For example, the addition of waste materials such as alum sludge can improve the California Bearing Ratio (CBR) and significantly improve the unconfined compressive strength (UCS) of clayey soils [10, 11]. This method is consistent with the circular economy, turning waste materials into useful and economical construction materials, while minimizing the environmental impacts of construction projects [12, 13].

The processes by which these materials enhance soil characteristics, including pozzolanic reactions, particle packing, and cation exchange, are complicated and [7]. Historically, the prediction of the resulting strength characteristics (i.e., UCS, CBR, shear strength) relies on significant and time-consuming laboratory work. But this is where Artificial Intelligence (AI) and Machine Learning (ML) can make a major difference. AI methods, particularly Artificial Neural Networks (ANNs), have been seen to model these nonlinear and complex relationships in geotechnical engineering quite well [14, 15]. There have been studies that illustrated ML algorithms can predict parameters, such as compressive strength, stress-strain behavior, and soil bearing capacity exceedingly well [10].

Nevertheless, while there has been a significant increase in the studies of both waste-based soil stabilization and the application of AI in geotechnics, research on the convergence of these two studies is still limited. Saad et al. [10] revealed that a systematic literature review focusing on the use of ML techniques to predict strength properties of soils that have been enhanced with green or waste materials has not been conducted. The current literature tends to treat ML in geotechnics in a general sense or consider the use of waste materials in isolation [16], but a synthesis of how AI can improve the efficiency of using these types of sustainable materials is critically missing. This paper therefore endeavors to provide a comprehensive review to fill this gap. The main objectives of this review are:

- To synthesize various literature sources that are the state-of-the-art of using a variety of waste materials to improve soil strength properties.
- To review and compare the performance of a variety of AI and ML models (such as ANN, extreme learning machine-ELM, group method of data handling-GMDH, hybrid algorithms) to predict the strength characteristics of waste-treated soils.
- To identify key challenges, such as data requirements, model selection, and interpretability, and to discuss future research directions for integrating AI into sustainable geotechnical design.

The objective of this review is to synthesize these two innovative fields and be a useful resource for researchers and practicing engineers interested in using AI to develop more efficient, cost-effective and environmentally conscious soil stabilization options.

6.2 Soil Stabilization Techniques and Methods

Soil stabilization is an important procedure of geotechnical and civil engineering, since improving the engineering properties of soil, including mechanical strength, permeability, compressibility, durability, and plasticity, is beneficial [17]. It is commonly utilized in subgrade constructions, embankments, road bases, foundations, and backfills for construction projects that require such (e.g., retaining walls and bridge abutments), among other areas of application. Stabilization methods are often grouped based on how physical, mechanical, and/or chemical methods can potentially address different soil types and project specifications [18]. Although the groupings are not always standardized, methods can be classified (to the maximum extent possible) in terms of their primary related mechanism of action whereby physical properties might be changed, chemical reaction initiated, or properties altered simultaneously by adding binders [17, 19]. The reviewed article discussed in detail the classification of methods as shown in Fig. 6.2, which generally indicates methods displayed on a scale from basic methods to advanced techniques [17]. The table demonstrates that soil stabilization methods can range from fundamental physical methods, such as excavating and replacing the existing material, to advanced chemical and mechanical-hydraulic methods of stabilization. Methods may be applied for either shallow, medium or deep stabilization and are often broadly classified into two method types:

6.2.1 *Physical Methods*

This area includes methods that change the soil's physical properties and structure usually without changing the underlying chemistry.

6.2.1.1 Electric and Thermal Methods

This category employs innovative processes, like Electrokinetic stabilization, which uses electricity to drive water and ions and compact and strengthen fine-grained soils. It also employs Heating or Vitrification, which applies high temperatures to dry out the soil, melt its constituents, and form a hard, ceramic-like mass [19]. These methods have a particularly strong application for certain situations, like in-situ treatment of contaminated soils or organic soils, such as peat, where conventional binders are less effective [20]. The methods ultimately aim for "Complete changes in soil physical properties," such as reduced permeability, greater density, and improved load-bearing characteristics as a result of the method.

6.2.1.2 Mechanical and Hydraulic Methods

Mechanical and hydraulic methods could be more accurately described as "mechanical and hydraulic improvement methods". These methods improve the properties of soil by physically rearranging and densifying soil particles and include (Fig. 6.1):

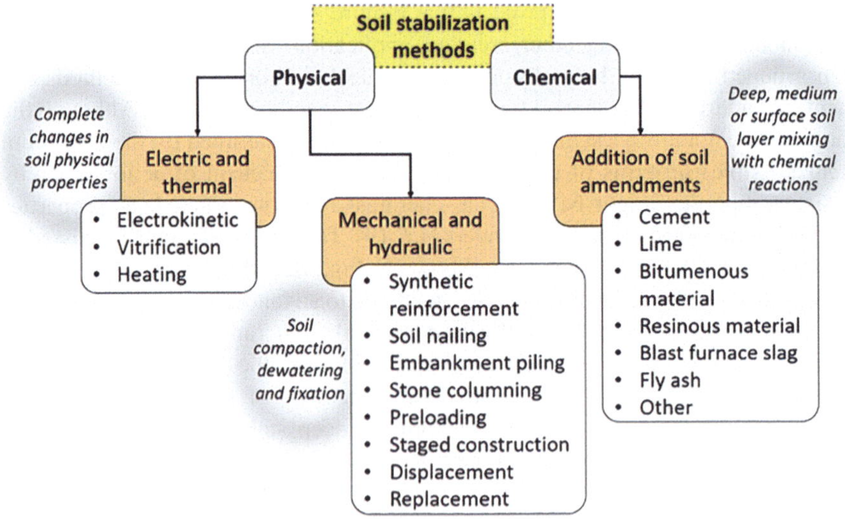

Fig. 6.1 General soil stabilization methods and their impact on soil [17]

- **Synthetic Reinforcement**: Use of geosynthetics (geogrids, geotextiles) to provide tensile strength [21–24].
- **Soil Compression, Dewatering, and Fixation**: Techniques like prefabricated vertical drains combined with surcharge (preloading) to accelerate consolidation.
- **Embankment Piling and Stone Columning**: Installing rigid inclusions (piles) or columns of granular material to improve load transfer and drainage [25, 26].
- **Preloading, Staged Construction, Displacement, Replacement**: Classical techniques involving applying loads to pre-consolidate soil, constructing in phases, or physically removing and replacing poor soil.

6.2.2 Chemical Methods: Addition of Soil Amendments

This particular section encapsulates the essence of Chemical Stabilization, annotated or addition of binders to the soil to initiate reactions that permanently improve its physical properties [27].

- **Conventional Primary Amendments**: More commonly traded binders include cement and lime [28, 29], and labels as Bituminous and Resinous Materials are binders that work by covering the soil grains with a thin layer of binder.
- **Sustainable Secondary Raw Materials (SRMs)**: Representing an actively adopted trend toward the circular economy, this classification importantly includes industrial by-products for construction materials, such as blast-furnace slag and fly ash [17, 27, 30]. Not infrequently as a result of pozzolanic or self-cementing properties, SRMs may replace some or all traditional binders, often resulting in reduced environmental impact from cement production [31]. The broadly referenced "Other" category in the SRM section incorporates newly emerging SRMs throughout the review as palm oil fuel ash (POFA), oil shale ash, pulp/paper mill ashes, as well as traditional adopted SRMs such as treated sawdust or wood powder, acrylonitrile or polylactic acid-based binding systems [32, 33].

Incorporating SRM into stabilization practices represents not only a technical solution but also a sustainability strategy to produce a new material while valuing waste by reducing GHG and conserving natural resources. These practices also benefit sustainability objectives, utilizing waste per action plan from circular economy, extending to global sustainability commitments [17].

6.3 Overview of Waste Materials for Soil Stabilization

Using industrial and agricultural waste materials to stabilize soils has developed to be a sustainable and technically feasible alternative to established stabilizers, including cement and lime. It is reported that a variety of waste materials, including industrial by-products, have been studied to potentially stabilize geotechnical properties in different soils. Industrial by-products include Fly Ash [30, 34–37], Waste Stone Powder (WSP) [38–42] and Brick Dust [43–46]. Agricultural materials, such as Rice Husk Ash (RHA) [1, 47–51] and Sugarcane Bagasse Ash (SCBA) [3, 52–55], are studied primarily for their high silica concentration. Other materials considered include Plastic, Jute, and various fibers [56–62], Eggshell Powder and Jaggery [63–66], Bitumen Emulsion [67], and various Chemical Additives [4, 68–70]. The selection of a waste material used in stabilization depends greatly on the soil type being stabilized, as seen in studies conducted on clayey, Lateritic, and Expansive soils [68, 71–73]. The wide variety of studies suggests that waste materials can provide effective economic and sustainable transformation in soil stabilization, while also addressing geotechnical stabilization and waste management.

Table 6.1 provides a summary of waste materials used in soil stabilization, including a few classifications, types of waste, notable references, and primary areas of application. Figure 6.2 presents a classification showing common waste materials used in soil stabilization.

6.4 AI Optimization Techniques in Geotechnical Engineering

Modern geotechnical engineering exists in an important intersection of civil infrastructure and earth sciences, where engineers engage with characterization of geomaterials (soil and rock) and exploring the complex behavior of these materials under static, dynamic, and hydraulic loads. The range of scope encompasses natural and engineered slope stability, shallow and deep foundation system performance and design, and ground-structure interaction of various structures (e.g., tunnels, retaining walls, embankments). The spatial variability and nonlinear, pressure dependence of these materials make advanced analysis necessary. This field has been relying even more on advanced artificial intelligence (AI) and machine learning (ML) methods to provide intelligent, model-based designs, and apply and develop advanced data-driven methods that rely on the inherent complexity and some ambiguity of these methods [15, 74–77].

These techniques are important for improving predictive accuracy, accelerating design, and contributing to sustainability by reducing dependence upon laborious, carbon-intensive laboratory and field testing. As illustrated in Fig. 6.3, which shows the proportionate distribution of the research areas in the field, there is an ample pedagogical emphasis on worth revisiting core Soil Properties (21%), Deep

Table 6.1 Categories of waste materials utilized in soil stabilization, classified by origin and efficacy

Category	Material	Reference	Tests	Notable advantages
Industrial by-products	Fly ash	[30, 34–37]	UCS, CBR	Pozzolanic properties improve strength and reduce plasticity, facilitating the utilization of large-volume waste.
	Waste stone powder (WSP)	[38–42]	UCS	Enhances compaction and stability; utilizes quarry/marble waste.
	Brick dust	[43–46]	UCS, direct shear, CBR	Improves shear strength and bearing capacity, while repurposing construction waste.
Agricultural waste	Rice husk ash (RHA)	[1, 47–51]	CBR, UCS	High silica content for pozzolanic reactions; effective in cohesive and expansive soils.
	Sugarcane bagasse ash (SCBA)	[3, 52–55]	CBR, UCS	Enhances strength and durability, making it ideal for lateritic and tropical soils.
	Corncob ash	[78, 79]	CBR, UCS	Improves the properties of expansive soils, such as black cotton soil.
Processed and synthetic waste	Scrap tire/ rubber	[80–84]	UCS, direct shear	Improves flexibility, reduces compaction, and enhances energy absorption, making it suitable for use in flexible pavements.
	Plastic/jute fibers	[56–62]	CBR	Provides tensile reinforcement, making it effective for erosion control and soil reinforcement.
Other non-traditional additives	Bitumen emulsion	[67]	CBR	Provides waterproofing and cohesion, making it suitable for sandy soils.
	Chemical additives	[4, 68–70]	UCS, CBR	Used for rapid strength gain and specialized stabilization needs.
	Egg shell powder, jaggery	[63–66]	UCS, consolidation	Alternative binders that can improve strength and settlement characteristics.

Foundations (11%), Rock Properties (10%), and Slope Stability (10%), but also a reasonable focus on Modeling (9%), Water-Soil Dynamics (3%), and Liquefaction (5%). Each of these focus areas has significant implications for simulation technologies, and will be significantly influenced by AI [74].

Core machine learning models, such as Artificial Neural Networks (ANNs), are recognized to estimate meaningful parameters focused on compression indices, shear strength, foundation settlements, and properties of rock including unconsolidated compressive strength (UCS) and Young's modulus, often to a higher level than conventional empirical strategies as they have the ability to utilize high-dimensional noise data without assumptions [85–88].

Fig. 6.2 Categorization and forms of common waste materials used in soil stabilization

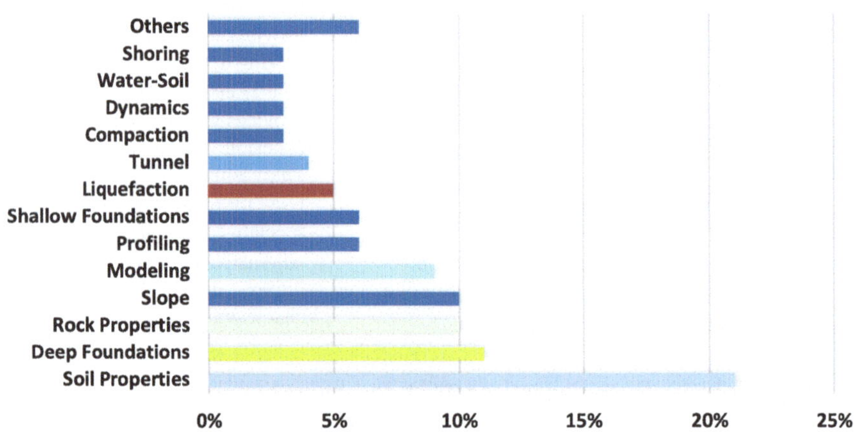

Fig. 6.3 Distribution of research focus areas in modern geotechnical engineering literature

Fuzzy Logic (FL) systems have effectively addressed uncertainty in site analysis and characterization, as well as slope stability [74, 89], and hybrid techniques such as the Adaptive Neuro-Fuzzy Inference System (ANFIS) integrate neural network adaptability with transparent and interpretable rule-based FL to assess associated properties and ground response, such as soil permeability and tunnel stability [90, 91]. In addition, Gene Expression Programming (GEP) creates explicit and accessible models for phenomena including soil liquefaction and blast vibrations [92, 93]. Ensemble methods, such as Random Forests (RF) and Support Vector Machines (SVM), further enhance prediction capabilities with advanced deep learning frameworks, such as Long Short-Term Memory networks (LSTMs) and Convolutional Neural Networks (CNNs) [15]. Similar improvements are observed for optimization problems, such as retaining wall and footing design, with inspiration from nature by utilizing metaheuristics, such as Particle Swarm Optimization (PSO) and Bacterial Foraging Optimization (BFO) [94]. In general, hybrid models based on PSO and ANN have been shown to improve accuracy for the prediction of mechanical properties of soil [95, 96]. The statistical technique of Analysis of Variance (ANOVA) can also be used to show the relative importance of factors designed to create materials [97]. Collectively, these artificial intelligence techniques provide a better way to manage risk and the opportunity for more knowledge-based decisions regarding slopes, tunnels, and foundations. Continued development of physics-guided AI and adaptive learning systems should lead to substantially improved interpretability and generalizability across geotechnical problems at different scales [98, 99], helping solidify the inclusion of AI as a central object in all modern geotechnical engineering projects.

6.5 Literature on AI Applications in Waste-Treated Soil Strength Assessment

The increasing focus on sustainability and waste valorization has brought a paradigm shift in geotechnical engineering, resulting in the use of industrial, agricultural, domestic, and mineral wastes as soil stabilizers. Materials are often classified based on origin: industrial wastes, agricultural wastes, domestic wastes, and mineral wastes. The use of these nonconventional, eco-friendly materials, especially agricultural wastes (e.g., rice husk ash—RHA), serves as a double benefit for the environment, as they enhance key geotechnical parameters such as unconfined compressive strength (UCS) and California Bearing Ratio (CBR) and provide an effective way of managing waste while helping to reduce the environmental impact of traditional stabilizers such as cement, the production of which is a key factor in CO_2 emissions [1, 100–102]. Although empirical models have been established to predict the strength of soils treated with certain waste types, the accuracy and application of these models applicable across different waste types can be limited. To overcome this, advanced computational methods, particularly Machine Learning (ML) are increasingly utilized to represent the inherent complexity and nonlinearity involved in soil stabilization in geotechnical engineering. The mechanical property prediction from stabilization options will be complex due to the nonlinear interactions of the soil medium and the additive when using waste material or stabilization using a waste material.

Soil stabilization using waste materials involves complicated nonlinear relationships between the soil matrix and additive; predicting mechanical properties such as UCS and CBR can therefore be complicated with traditional statistical means. Artificial Intelligence (AI) is becoming an important and powerful approach for modeling these relationships. Besides the commonly used Artificial Neural Networks (ANNs), which are often referred to as "black-box" models due to their general lack of interpretability, more recent studies have proposed more interpretable "grey-box" models such as Genetic Programming (GP) or Classification and Regression Trees (CART), which provide a trade-off between predictive accuracy and transparency, because GPs provide explicit predictions as mathematical equations, and CARTs provide explicit predictions as decision trees, which geotechnical engineers can use [103, 104]. AI techniques are now used in many different geotechnical applications, such as analysis of frozen soils [105], rock mechanics [106], slope stability [107], and soil dynamics [14], which shows the versatility of AI techniques.

A pioneering study by Yao et al. [108] established a hybrid TPE-XGB model for predicting the unconfined compressive strength (UCS) of soil treated with waste. With a database of 474 specimens that included chemical compositions, their model averaged significantly better than other machine learning models. The SHAP analysis revealed curing age, cement content, and the Al_2O_3/SiO_2 in waste contributed the most, with 10% waste being the optimum. This artificial intelligence framework provides a robust framework for improving waste valorization in sustainable construction.

In conjunction with this, Onyelowe et al. [109] undertook an extensive comparison of nine machine learning models to predict the UCS of cohesive soil stabilized with two commonly used industrial by-products, lime and cement. Their research employed a significant database, which had 190 experiments. The outcomes showed a wide divergence of performance among the models. The K-NN and GB models provided the most accurate predictive performance (95%). A distinguishing feature of their research involved a robust sensitivity analysis, which indicated that the Maximum Dry Density (MDD) was the driving parameter affecting UCS, with the soil consistency limits and cement content following closely after it.

Building on the diversity of waste materials studied, Aamir et al. [71] focused on the performance of alum sludge (AS), a common waste byproduct in water treatment facilities, as a soil stabilizer for weak, clayey soil (classified A-7-6 in AASHTO and CH in USCS). Their experimental investigation involved mixing dried and crushed AS with the soil at percentages from 0% to 10% by dry weight. Results showed an optimal dose: the CBR of the soil significantly increased with AS addition, reaching a maximum value of 15.75% CBR with 8% AS, which was more than three times higher than that of the untreated soil. At that stage, an additional increase to 10% AS caused the CBR value to reduce to 7.54%, demonstrating that more than optimal AS can diminish strength, but the CBR was still greater than the untreated soil. The authors also noted that AS lowered the Optimum Moisture Content (OMC) of the soil, offering potential energy savings in field compaction. To capture this complexity, the authors also developed an Artificial Neural Network (ANN) model to predict CBR based on the four input parameters AS, PI, OMC, and OMC.

Similarly, Priyadarshee et al. [110] implemented an Artificial Neural Network (ANN) with a 5-10-1 structure to make predictions regarding the UCS of Kaolin clay modified with pond ash, rice husk ash, and cement. Their model had curing period and admixture percentages as input, and recorded better prediction accuracy ($R^2 = 0.98$) than their Multiple Regression Analysis model ($R^2 = 0.88$). The study also included a sensitivity analysis that showed the waste admixtures together contributed 70% to the strength development, illustrating how effective ANN is for modeling complex soil stabilization.

Expanding on the use of agricultural waste, Baghbani et al. [103] conducted a novel study to predict the strength of expansive soil stabilized with Hydrated Lime Activated Rice Husk Ash (HARHA). They presented two grey-box machine learning models, CART and GP, to predict CBR, UCS, and resistance value (R-value). The GP model gave explicit equations ready for applications while the CART model produced decision trees that were interpretable. The GP model performed better than both methods, particularly for UCS and R-value predictions, showing a greatly high coefficient of determination ($R^2 > 0.99$). A significant contribution of this study was the sensitivity analysis that ascertained HARHA content, liquid limit (LL), and maximum dry density (MDD) were the most significant input parameters influencing strength. The research indicates that the use of transparent AI models may allow for accurate predictions and actionable recommendations for geotechnical designs involving waste-modified soils.

Additionally, it is worth noting the study undertaken by Mohammed et al. [111], which supports that Ground-Granulated Blast-Furnace Slag (GGBS) increases the UCS of soft soils, and produced an XGBoost model ($R^2 = 0.95$) that similarly demonstrated good accuracy; however, it had to consider a small dataset size and the slow hydration rate of GGBS. Topçuoglu [112] found that fly ash and a relatively small amount of blast-furnace slag significantly improved kaolin clay strength ($q_u = 1641.6$ kPa) and a proposed ANN model provided good predictive capability ($R^2 = 0.97$). Shariatmadari et al. [113] considered volcanic ash-slag blends for sandy soils and reported UCS values up to 6218.6 kPa where the ANN ($R^2 = 0.97$) performed better than an evolutionary polynomial regression model. Mozumder and Laskar [114] valued the blended GGBS and fly ash for clayey soils and the ANN ($R^2 = 0.98$) model was produced for superior accuracy against regression models but required accurate chemical proportions. Eyo et al. [115] assessed various agricultural ashes (RHA, POFA, bagasse, corn cob, etc.) and indicated that classification (Accuracy = 0.92) and regression ($R^2 = 0.90$) models provided reliable UCS predictions.

Finally, Tabarsa et al. [116] emphasized the combined effect of RHA together with cement or lime on silty and sandy soils, where utilizing SVM-RBF ($R^2 = 0.997$) produced the best results compared to the other approaches.

Overall, this review has provided evidence of the groundbreaking influence of artificial intelligence on enhancing the predictive accuracy and operational utility, within an engineering context, of waste-amended soil stabilization. As presented in Table 6.2, studies have consistently demonstrated the predictive power of machine learning models, including ANNs, SVMs, and more recent ensemble and hybrid

Table 6.2 Summary of AI model performance in predicting the strength of soils treated with various waste materials

Reference	Waste type	Soil type	Content (%)	Strength improvement	AI model performance	Limitations
[111]	GGBS	Soft soils (cohesive, expansive, peat)	0–50	Significant UCS improvement	XGBoost: $R^2 = 0.95$, MAE = 0.049, RMSE = 0.077	Limited dataset ($n = 200$), slow hydration of GGBS, reliance on XAI methods
[112]	Fly ash + 1% BF	Kaolin clay (CL)	1% BF + 9% FA	Max $q_u = 1641.6$ kPa at 56 days	ANN: $R^2 = 0.97$, RMSE = 0.04–0.10	Restricted to kaolin clay, fixed BF content, no microstructural analysis, small dataset
[103]	Rice husk ash (HARHA)	Expansive clay	0.1–12	CBR, UCS, R-value ↑ up to ~11.5%	CART: $R^2 = 0.975$–0.990; GP: $R^2 = 0.990$–0.998	Lab-specific data, computational complexity, and overfitting risk
[71]	Alum sludge (AS)	A-7-6 (CH)	0–10 (opt. 8)	CBR ↑ from 5.35% to 15.75% (~3×)	ANN: $R^2 \approx 1.0$, RMSE = 0.058	Decline beyond 8%, health/environmental concerns, no long-term study
[109]	Cement and lime	Cohesive soil	0–30	UCS >5.4 MPa	GB, K-NN: $R^2 = 0.99$, Acc = 95%	Black-box models, range-limited validity, no simple predictive formula
[117]	Fly ash (C&F)	Cohesive clay (CH)	5–25	UCS1 ↑ 547%, UCS28 ↑ 893%	ANN: $R^2 = 0.64$–0.83, RMSE = 29–49	Black-box nature, cohesive soils only, requires lab inputs
[113]	VA + GGBFS	Poorly graded sand (SP)	3–10	UCS ↑ to 6218.6 kPa	ANN: $R^2 = 0.97$, RMSE = 0.044, superior to EPR	Requires complex chemical ratios, sandy soils only, and practical complexity
[118]	FA + BFS	Clayey soils	FA: 0–20, BFS: 0–50	UCS ↑ up to 24.26 MPa	NF-GMDH-PSO: $R^2 = 0.94$, MAE = 0.83	Valid for UCS <10 MPa, soil-specific, needs further validation

Reference	Waste type	Soil type	Content (%)	Strength improvement	AI model performance	Limitations
[114]	GGBS + FA	Clayey soils (CH, CL)	GGBS: 4–50, FA: 4–20	UCS ↑ to 24.26 MPa	ANN: $R = 0.982$, MAPE = 8.34%	Dataset-specific, black-box model requires detailed inputs and cures within 28 days
[11]	Alum sludge	Clay (CH → ML)	0–10 (opt. 8)	CBR ↑ 2–3× at different energies	ANN: $R^2 = 0.97$–0.999, RMSE = 0.02–0.62	One soil type, lab-scale only, no durability/leachate study
[115]	Multiple agricultural wastes (RHA, POFA, etc.)	Clayey soils (CL, CH, MH)	Pozzolan: 0.1–25; Additive: 0.5–11.25	UCS improved	Classification: Acc = 0.92, Precision = 0.94; Regression: $R^2 = 0.90$, MAPE = 0.34	Lab-only, no durability/settlement modeling, field validation required
[116]	RHA + cement/lime	SM, MH	Cement: 1.25–3.75; Lime: 2.5–7.5; RHA: 0–3.75	Significant UCS gain (CLR > CL)	SVM-RBF: $R = 0.997$, AAPE = 4.6%; ANN: $R > 0.995$	Data-specific, generalizability limited, sensitive to model parameters

models, in capturing nonlinear, complex relationships between waste type, soil properties, and related improvement in mechanical behavior. Some clear trends in the literature include the importance of maximum dry density (MDD) and the recommended dosage of waste materials including alum sludge, fly ash, and rice husk ash. While these models were promising, it did also note limitations surrounding TAM and transferability, dataset and access to data limitations, and need for validation in-field. Future work should aim to further build multi-scale databases, improve transparency via XAI, and assess long-term environmental and durability aspects of waste-amended soil stabilization, to promote sustainability and data-driven geotechnical design.

6.6 Conclusions

This review explores the convergence between two important areas of geotechnical engineering: the use of sustainable waste materials for soil improvement and the development of Artificial Intelligence (AI)-based models to predict the strength properties resulting from the use of waste materials. The analysis of the current literature has resulted in the following main conclusions:

1. A wide range of industrial, agricultural, domestic, and mineral by-products have been demonstrated to enhance the performance of problematic soils from an engineering perspective. In many cases, those materials include materials such as Cement Kiln Dust (CKD), fly ash, rice husk ash (RHA), ground-granulated blast-furnace slag (GGBS), and marble dust, which can enhance key engineering performance framework considerations, UCS and CBR. The mechanics of these waste materials interacting with soil are complex and nonlinear, meaning the results depend on many factors including soil type, wastes, and portion of waste to soil. The proportions of waste are generally in the range of 2–50% by weight of the total considering soil plus waste.
2. The use of traditional empirical and statistical methodologies often does not help to model the value of the complex relationship of waste-amended soils as evidenced in many studies. AI and machine learning (ML) methodologies such as Artificial Neural Networks (ANN), Support Vector Machines (SVM), Genetic Programming (GP), and ensemble models such as XGBoost have been shown to be dramatically superior. The AI models have returned very high prediction accuracy levels (commonly R^2 values >0.95) in assessing strength parameters of UCS and CBR, outperforming more traditional regression analyses.
3. AI models have been shown to be highly effective in identifying the key parameters that influence their strength development. They employ advanced sensitivity analyses (e.g., SHAP, feature importance) to demonstrate that curing age, maximum dry density (MDD), chemical composition of the waste (e.g., alumina

and silica content), and ideal amount of additive are always identified as key parameters. This information is very helpful in designing the mixture and can lead to justification of experimental work with the materials under investigation.

4. The review addresses a significant research niche in the literature, providing a systematic synthesis of the existing literature on sustainable soil stabilization/disposal utilization and AI. While there are extensive studies related to each topic alone, this synthesis examines the evidence on how AI can help optimize the potential efficiencies, economies, and beneficial impacts on the environment of waste materials (and the circular economy principles) in construction.

6.7 Challenges and Future Directions

Although the findings reported here are promising, challenges will remain. Future work should concentrate on:

- Developing large, consistent, open-access datasets that incorporate much more diversity of soil, waste types, and long-term performance data (e.g., durability to environmental stressors).
- Transitioning from "black-box" models and improving Explainable AI (XAI) and other gray-box type models (e.g., GP) that would allow for transparent, action-oriented formulas for users.
- Transitioning beyond lab-based predictive modeling to field-based applications that validate AI under predictive processes that is not land-waste-based, and ultimately understanding AI incorporation into geotechnical code and practice.
- Link AI strength predictive models to LCA tools measuring all mechanical performance measures and integrating environmental and economic benefits of alternative waste-to-soil solutions.

In summary, the integration of artificial intelligence into predictive modeling of material performance and the utilization of sustainable materials as waste in engineered soil is a transformation in practice in geotechnical engineering. The integrated nature of these AI-based models and sustainable waste materials provides significant pathways to waste management issues, while advancing safer, less expensive, environmentally sustainable infrastructure while creating a pathway to smarter and more sustainable construction.

AI Use Disclosure Statement The author has only used Grammarly for language improvement and proofreading in the preparation and revision of this manuscript.

The author used AI-assisted engineering tools such as SCISPACE and Elicit to search and gather relevant academic materials.

References

1. Rahgozar MA, Saberian M, Li J (2018) Soil stabilization with non-conventional eco-friendly agricultural waste materials: an experimental study. Transp Geotech 14:52–60. https://doi.org/10.1016/j.trgeo.2017.09.004
2. Utkarsh, Jain PK (2024) Enhancing the properties of swelling soils with lime, fly ash, and expanded polystyrene—a review. Heliyon 10(12):e32908. https://doi.org/10.1016/j.heliyon.2024.e32908
3. Adnan M, Kumar S, Garg N, Kumar Gupta K, Kumar Das S (2023) Soil stabilization using waste "Bagasse ash and lime": a review. Mater Today Proc. https://doi.org/10.1016/j.matpr.2023.05.018
4. Barman D, Dash SK (2022) Stabilization of expansive soils using chemical additives: a review. J Rock Mech Geotech Eng 14(4):1319–1342. https://doi.org/10.1016/j.jrmge.2022.02.011
5. Barbhuiya S, Kanavaris F, Das BB, Idrees M (2024) Decarbonising cement and concrete production: strategies, challenges and pathways for sustainable development. J Build Eng 86:108861. https://doi.org/10.1016/j.jobe.2024.108861
6. Behnood A (2018) Soil and clay stabilization with calcium- and non-calcium-based additives: a state-of-the-art review of challenges, approaches and techniques. Transp Geotech 17:14–32. https://doi.org/10.1016/j.trgeo.2018.08.002
7. Shinde B, Sangale A, Pranita M, Sanagle J, Roham C (2024) Utilization of waste materials for soil stabilization: a comprehensive review. Prog Eng Sci 1(2):100009. https://doi.org/10.1016/j.pes.2024.100009
8. Al-Taie AJ, Al-Obaidi A, Alzuhairi M (2020) Utilization of depolymerized recycled polyethylene terephthalate in improving poorly graded soil. Transp Infrastruct Geotechnol 7(2):206–223. https://doi.org/10.1007/s40515-019-00099-2
9. Blayi RA, Sherwani AFH, Ibrahim HH, Faraj RH, Daraei A (2020) Strength improvement of expansive soil by utilizing waste glass powder. Case Stud Constr Mater 13:e00427. https://doi.org/10.1016/j.cscm.2020.e00427
10. Saad AH et al (2023) A systematic review of machine learning techniques and applications in soil improvement using green materials. Sustainability 15(12):9738. https://doi.org/10.3390/su15129738
11. Shah SAR, Mahmood Z, Nisar A, Aamir M, Farid A, Waseem M (2020) Compaction performance analysis of alum sludge waste modified soil. Constr Build Mater 230:116953. https://doi.org/10.1016/j.conbuildmat.2019.116953
12. Abbaspour M, Aflaki E, Moghadas Nejad F (2019) Reuse of waste tire textile fibers as soil reinforcement. J Clean Prod 207:1059–1071. https://doi.org/10.1016/j.jclepro.2018.09.253
13. Zaini MSI, Hasan M (2023) Effect of optimum utilization of silica fume and lime on the stabilization of problematic soils. Int J Integr Eng 15(1):352–366
14. Baghbani A, Choudhury T, Costa S, Reiner J (2022) Application of artificial intelligence in geotechnical engineering: a state-of-the-art review. Earth Sci Rev 228:103991. https://doi.org/10.1016/j.earscirev.2022.103991
15. Liu H, Su H, Sun L, Dias-da-Costa D (2024) State-of-the-art review on the use of AI-enhanced computational mechanics in geotechnical engineering. Artif Intell Rev 57(8):196. https://doi.org/10.1007/s10462-024-10836-w
16. Moayedi H, Mosallanezhad M, Rashid ASA, Jusoh WAW, Muazu MA (2020) A systematic review and meta-analysis of artificial neural network application in geotechnical engineering: theory and applications. Neural Comput Appl 32(2):495–518. https://doi.org/10.1007/s00521-019-04109-9
17. Vincevica-Gaile Z et al (2021) Towards sustainable soil stabilization in peatlands: secondary raw materials as an alternative. Sustainability 13(12):6726. https://doi.org/10.3390/su13126726

18. Hejazi SM, Sheikhzadeh M, Abtahi SM, Zadhoush A (2012) A simple review of soil reinforcement by using natural and synthetic fibers. Constr Build Mater 30:100–116. https://doi.org/10.1016/j.conbuildmat.2011.11.045
19. Guillaud H et al (1995) Compressed earth blocks: manual of design and construction. https://www.semanticscholar.org/paper/COMPRESSED-EARTH-BLOCKS%3A-MANUAL-OF-DESIGN-AND-Guillaud-Joffroy/46e1eaba30d6ce8e9af1cf98574fa542717c926d. Accessed 6 Sept 2025
20. Gomes Correia A, Winter MG, Puppala AJ (2016) A review of sustainable approaches in transport infrastructure geotechnics. Transp Geotech 7:21–28. https://doi.org/10.1016/j.trgeo.2016.03.003
21. Abdelkader B, Arab A, Sadek M, Shahrour I (2018) Laboratory investigation of the influence of geotextile on the stress–strain and volumetric change behavior of sand. Geotech Geol Eng 36(4):2077–2085. https://doi.org/10.1007/s10706-018-0446-6
22. Afzali-Nejad A, Lashkari A, Shourijeh PT (2017) Influence of particle shape on the shear strength and dilation of sand-woven geotextile interfaces. Geotext Geomembr 45(1):54–66. https://doi.org/10.1016/j.geotexmem.2016.07.005
23. Altun S, Göktepe AB, Lav MA (2008) Liquefaction resistance of sand reinforced with geosynthetics. Geosynth Int 15(5):322–332. https://doi.org/10.1680/gein.2008.15.5.322
24. Dash SK, Rajagopal K, Krishnaswamy NR (2004) Performance of different geosynthetic reinforcement materials in sand foundations. Geosynth Int 11(1):35–42. https://doi.org/10.1680/gein.2004.11.1.35
25. Ghorbani A, Hosseinpour I, Shormage M (2021) Deformation and stability analysis of embankment over stone column-strengthened soft ground. KSCE J Civil Eng 25(2):404–416. https://doi.org/10.1007/s12205-020-0349-y
26. Rosidi D, Gharabegian A, Jang D-J (2012) Dynamic response of an embankment during stone-column piling, pp 1–10. https://doi.org/10.1061/40975(318)105
27. Sharma AK, Sivapullaiah PV (2016) Ground-granulated blast-furnace slag amended fly ash as an expansive soil stabilizer. Soils Found 56(2):205–212. https://doi.org/10.1016/j.sandf.2016.02.004
28. Al-Rawas AA, Hago AW, Al-Sarmi H (2005) Effect of lime, cement and Sarooj (artificial pozzolan) on the swelling potential of an expansive soil from Oman. Build Environ 40(5):681–687. https://doi.org/10.1016/j.buildenv.2004.08.028
29. Senneca O et al (2020) Mechanisms affecting the delayed efficiency of cement-based stabilization/solidification processes. J Clean Prod 261:121230. https://doi.org/10.1016/j.jclepro.2020.121230
30. Abdila SR et al (2022) Potential of soil stabilization using ground-granulated blast-furnace slag (GGBFS) and fly ash via geopolymerization method: a review. Materials 15(1):375. https://doi.org/10.3390/ma15010375
31. Juha F, Leena K-T, Pyry P, Juha F, Leena K-T, Pyry P (2018) Mass stabilization as a ground improvement method for soft peaty. In: Peat. IntechOpen. https://doi.org/10.5772/intechopen.74144
32. Cherian C, Siddiqua S (2019) Pulp and paper mill fly ash: a review. Sustainability 11(16):4394. https://doi.org/10.3390/su11164394
33. Mahmood AA, Hussain MK, Mohamad SNA (2020) Use of palm oil fuel ash (POFA)-stabilized Sarawak peat composite for road subbase. Mater Today Proc 20:505–511. https://doi.org/10.1016/j.matpr.2019.09.178
34. Andavan S, Pagadala VK (2020) Experimental study on addition of lime and fly ash for the soil stabilization. Mater Today Proc 22:1065–1069. https://doi.org/10.1016/j.matpr.2019.11.300
35. Baloochi H, Aponte D, Barra M (2020) Soil stabilization using waste paper fly ash: precautions for its correct use. Appl Sci 10(23):8750. https://doi.org/10.3390/app10238750
36. Noaman MF, Khan MA, Ali K, Hassan A (2022) A review on the effect of fly ash on the geotechnical properties and stability of soil. Clean Mater 6:100151. https://doi.org/10.1016/j.clema.2022.100151

37. Vukićević M, Pujević V, Marjanović M, Jocković S, Maraš-Dragojević S (2015) Fine grained soil stabilization using class F fly ash with and without cement. In: Winter MG, Smith DM, Eldred PJL, Toll DG (eds) Geotechnical engineering for infrastructure and development: XVI European conference on soil mechanics and geotechnical engineering, vol 1–7. https://doi.org/10.1680/ecsmge.60678.vol5.413
38. Abdelkader HAM, Hussein MMA, Ye H (2021) Influence of waste marble dust on the improvement of expansive clay soils. Adv Civil Eng 2021(1):3192122. https://doi.org/10.1155/2021/3192122
39. Li J et al (2024) Utilizing waste stone powder for improving properties of phosphogypsum-based composite prepared by semi-dry method. Constr Build Mater 426:136095. https://doi.org/10.1016/j.conbuildmat.2024.136095
40. Niraula U, Dahal BK, Acharya S, Phuyal P (2025) High-plasticity silt stabilization: role of waste stone dust, cement, and curing time. Results Eng 26:104877. https://doi.org/10.1016/j.rineng.2025.104877
41. Roohbakhshan A, Kalantari B (2014) Effect of lime and waste stone powder variation on the pH values, moisture content and dry density of clayey soil. Int J Adv Appl Sci 3(1):41–46. https://doi.org/10.11591/ijaas.v3.i1.pp41-46
42. Umar IH, Lin H (2024) Marble powder as a soil stabilizer: an experimental investigation of the geotechnical properties and unconfined compressive strength analysis. Materials 17(5):1208. https://doi.org/10.3390/ma17051208
43. Kasinikota P, Tripura DD (2021) Evaluation of compressed stabilized earth block properties using crushed brick waste. Constr Build Mater 280:122520. https://doi.org/10.1016/j.conbuildmat.2021.122520
44. Kongkajun N, Laitila EA, Ineure P, Prakaypan W, Cherdhirunkorn B, Chakartnarodom P (2020) Soil-cement bricks produced from local clay brick waste and soft sludge from fiber cement production. Case Stud Constr Mater 13:e00448. https://doi.org/10.1016/j.cscm.2020.e00448
45. Kumar V, Priyadarshee A, Rahul AK, Kumar S, Kumar V, Kumar A (2025) Evaluating the influence of cement addition on the geotechnical properties of brick dust–mixed local soil. Procedia Struct Integr 70:303–310. https://doi.org/10.1016/j.prostr.2025.07.057
46. Oti JE, Kinuthia JM, Robinson RB (2014) The development of unfired clay building material using brick dust waste and Mercia mudstone clay. Appl Clay Sci 102:148–154. https://doi.org/10.1016/j.clay.2014.09.031
47. Basha EA, Hashim R, Mahmud HB, Muntohar AS (2005) Stabilization of residual soil with rice husk ash and cement. Constr Build Mater 19(6):448–453. https://doi.org/10.1016/j.conbuildmat.2004.08.001
48. Nasiri M, Lotfalian M, Modarres A, Wu W (2017) Use of rice husk ash as a stabilizer to reduce soil loss and runoff rates on sub-base materials of forest roads from rainfall simulation tests. Catena 150:116–123. https://doi.org/10.1016/j.catena.2016.11.010
49. Raja K, Venkatachalam S, Vishnuvardhan K, Krishnan RSR, Selvan VT, Vetriselvan N (2022) A review on soil stabilization using rice husk ash and lime sludge. Mater Today Proc 65:1205–1212. https://doi.org/10.1016/j.matpr.2022.04.178
50. Shah MM, Li H (2025) Utilization of lime-mixed geopolymerized rice husk ash (LGR) for balanced amelioration of collapsible soil. J Rock Mech Geotech Eng 17(6):3925–3949. https://doi.org/10.1016/j.jrmge.2024.12.022
51. Uma Maguesvari M, Muthaiyan P, Yugasini S (2023) Stabilization of soil using kenaf fiber and rice husk ash. Mater Today Proc. https://doi.org/10.1016/j.matpr.2023.06.261
52. Azhar MB, Mujtaba H, Farooq K, Shah MM, Masoud Z (2025) Bridging experimental work and environmental application: LCA-driven scalability of bagasse ash for eco-friendly stabilization of swelling clay subgrade. Ain Shams Eng J 16(10):103623. https://doi.org/10.1016/j.asej.2025.103623

53. Moazami A, Nabizadeh M, Komaei A, Soroush A (2025) Sustainable soil stabilization with sodium carbonate-activated slag, pumice, rice husk, and sugarcane bagasse ashes: a life cycle assessment approach. Results Eng 27:106900. https://doi.org/10.1016/j.rineng.2025.106900
54. Paul S, Chakraborty P, Anwar AB, Mortoza G, Sukti LS (2025) Effect of cement and sugarcane bagasse ash on the micro-mechanical characteristics of expansive soil. Results Eng 27:106874. https://doi.org/10.1016/j.rineng.2025.106874
55. Singh S, Chohan JS, Kumar R, Gupta PK (2022) Stability of compressed earth blocks using sugarcane bagasse ash and wheat straw. Mater Today Proc 51:993–997. https://doi.org/10.1016/j.matpr.2021.07.023
56. Boobalan SC, Anandakumar PK, Sathasivam M (2023) Utilization of waste plastic sheets as soil stabilization materials. Mater Today Proc. https://doi.org/10.1016/j.matpr.2023.07.184
57. Gangwar P, Tiwari S (2021) Stabilization of soil with waste plastic bottles. Mater Today Proc 47:3802–3806. https://doi.org/10.1016/j.matpr.2021.03.010
58. Güllü H, Khudir A (2014) Effect of freeze–thaw cycles on unconfined compressive strength of fine-grained soil treated with jute fiber, steel fiber and lime. Cold Reg Sci Technol 106–107:55–65. https://doi.org/10.1016/j.coldregions.2014.06.008
59. Gupta A, Kumar M (2022) Clayey soil stabilization using fly ash and jute fiber. Mater Today Proc 48:1205–1210. https://doi.org/10.1016/j.matpr.2021.08.246
60. Hatefi MH, Arabani M, Payan M, Ranjbar PZ (2025) Impact of recycled macro-synthetic fibers on kaolin soil stabilized with lime kiln dust and volcanic ash. Case Stud Constr Mater 22:e04830. https://doi.org/10.1016/j.cscm.2025.e04830
61. Magray AR, Singh P (2023) Improvement in the properties of soil by utilization of jute fiber with black carbon as a soil reinforcement. Mater Today Proc. https://doi.org/10.1016/j.matpr.2023.04.024
62. Paul S, Tasnim A, Majumder J (2025) Enhancement of micro-mechanical characteristics of expansive soil through synergistic incorporation of jute and nylon fibers with cement. Results Eng 26:104685. https://doi.org/10.1016/j.rineng.2025.104685
63. Anburuvel A, Sathiparan N, Dhananjaya GMA, Anuruththan A (2023) Characteristic evaluation of geopolymer based lateritic soil stabilization enriched with eggshell ash and rice husk ash for road construction: an experimental investigation. Constr Build Mater 387:131659. https://doi.org/10.1016/j.conbuildmat.2023.131659
64. Elavarasan S, Priya AK, Elango KS, Shahithya S (2023) A review on addition of egg shell powder to CSEB blocks. Mater Today Proc. https://doi.org/10.1016/j.matpr.2023.03.718
65. Saldanha RB, da Rocha CG, Caicedo AML, Consoli NC (2021) Technical and environmental performance of eggshell lime for soil stabilization. Constr Build Mater 298:123648. https://doi.org/10.1016/j.conbuildmat.2021.123648
66. Yang B, Yin Z, Hasheminezhad A, Alsheyab MA, Ceylan H, Kim S (2025) Enhancing soil engineering properties with eggshell powder: a laboratory-based approach to sustainable road construction. J Clean Prod 505:145464. https://doi.org/10.1016/j.jclepro.2025.145464
67. Andavan S, Kumar BM (2020) Case study on soil stabilization by using bitumen emulsions—a review. Mater Today Proc 22:1200–1202. https://doi.org/10.1016/j.matpr.2019.12.121
68. Almajed A, Abbas H, Arab M, Alsabhan A, Hamid W, Al-Salloum Y (2020) Enzyme-induced carbonate precipitation (EICP)-based methods for eco-friendly stabilization of different types of natural sands. J Clean Prod 274:122627. https://doi.org/10.1016/j.jclepro.2020.122627
69. Puppala AJ (2016) Advances in ground modification with chemical additives: from theory to practice. Transp Geotech 9:123–138. https://doi.org/10.1016/j.trgeo.2016.08.004
70. Rangaswamy K, Mohan RP (2025) Strengthening the subgrade clay soil using nano chemical stabilization. Soils Found 65(4):101638. https://doi.org/10.1016/j.sandf.2025.101638
71. Aamir M et al (2019) Performance evaluation of sustainable soil stabilization process using waste materials. Processes 7(6):378. https://doi.org/10.3390/pr7060378
72. Gidebo FA, Yasuhara H, Kinoshita N (2023) Stabilization of expansive soil with agricultural waste additives: a review. Int J Geoeng 14(1):14. https://doi.org/10.1186/s40703-023-00194-x

73. Ojuri OO et al (2022) Eco-friendly stabilization of highway lateritic soil with cow bone powder admixed lime and plastic granules reinforcement. Clean Waste Syst 2:100012. https://doi.org/10.1016/j.clwas.2022.100012
74. Ebid AM (2021) 35 Years of (AI) in geotechnical engineering: state of the art. Geotech Geol Eng 39(2):637–690. https://doi.org/10.1007/s10706-020-01536-7
75. Onyelowe KC et al (2023) Selected AI optimization techniques and applications in geotechnical engineering. Cogent Eng 10(1):2153419. https://doi.org/10.1080/23311916.2022.2153419
76. Armaghani, D. J., Mohamad, E. T., Narayanasamy, M. S., Narita, N., & Yagiz, S. (2017). Development of hybrid intelligent models for predicting TBM penetration rate in hard rock condition. Tunnelling and Underground Space Technology, 63, 29–43
77. Yong, W., Zhou, J., Jahed Armaghani, D., Tahir, M. M., Tarinejad, R., Pham, B. T., & Van Huynh, V. (2021). A new hybrid simulated annealing-based genetic programming technique to predict the ultimate bearing capacity of piles. Engineering with computers, 37(3), 2111–2127
78. Singh S (2021) Experimental investigation of corn cob ash on silty clay stabilized with calcium carbide. Mater Today Proc 37:3658–3660. https://doi.org/10.1016/j.matpr.2020.10.146
79. Yifru W, Getu N, Kifile D, Mesfin A, Sewunet A, Tamene M (2022) Effects of corn cob ash as partial replacement of cement for stabilization of an expansive clay. Adv Civil Eng 2022(1):6788120. https://doi.org/10.1155/2022/6788120
80. Al-Subari L, Ekinci A (2022) Strength and durability assessment of shredded tire rubber stabilized artificially cemented alluvial clay. Constr Build Mater 345:128312. https://doi.org/10.1016/j.conbuildmat.2022.128312
81. Bekhiti M, Ghrieb A, Smaida A (2021) Influence of worn tire rubber fibers on the swelling potential and pressure of clay soils. Mater Today Proc 45:5225–5230. https://doi.org/10.1016/j.matpr.2021.01.723
82. Groenevelt PH, Grunthal PE (1998) Utilization of crumb rubber as a soil amendment for sports turf. Soil Tillage Res 47(1):169–172. https://doi.org/10.1016/S0167-1987(98)00089-0
83. Kavinkumar V, Elavarasan S, Anandaraj S, Dharmaraj R (2021) Development of subgrade layer using scrap tire and stabilized expansive soil. Mater Today Proc 37:986–991. https://doi.org/10.1016/j.matpr.2020.06.188
84. Kumar M (2021) Clay soil stabilization by utilizing secondary lime and rubber tire powder. Mater Today Proc 37:3471–3474. https://doi.org/10.1016/j.matpr.2020.09.383
85. Jan MS et al (2023) Appraisal of different artificial intelligence techniques for the prediction of marble strength. Sustainability 15(11):8835. https://doi.org/10.3390/su15118835
86. Khan NM et al (2022) Application of machine learning and multivariate statistics to predict uniaxial compressive strength and static Young's modulus using physical properties under different thermal conditions. Sustainability 14(16):9901. https://doi.org/10.3390/su14169901
87. Luo T, Wang J, Chen L, Sun C, Liu Q, Wang F (2024) Quantitative characterization of the brittleness of deep shales by integrating mineral content, elastic parameters, in situ stress conditions and logging analysis. Int J Coal Sci Technol 11(1):10. https://doi.org/10.1007/s40789-023-00637-x
88. Shahin MA, Maier HR, Jaksa MB (2002) Predicting settlement of shallow foundations using neural networks. J Geotech Geoenviron Eng 128(9):785–793. https://doi.org/10.1061/(ASCE)1090-0241(2002)128:9(785)
89. Mohamed T, Kasa A, Taha MR (2012) Fuzzy logic system for slope stability prediction. Int J Adv Sci Eng Inf Technol 2(2):151–155. https://doi.org/10.18517/ijaseit.2.2.174
90. Cabalar AF, Cevik A, Gokceoglu C (2012) Some applications of adaptive neuro-fuzzy inference system (ANFIS) in geotechnical engineering. Comput Geotech 40:14–33. https://doi.org/10.1016/j.compgeo.2011.09.008
91. Kalkan E, Akbulut S, Tortum A, Celik S (2009) Prediction of the unconfined compressive strength of compacted granular soils by using inference systems. Environ Geol 58(7):1429–1440. https://doi.org/10.1007/s00254-008-1645-x
92. Goharzay M, Noorzad A, Ardakani AM, Jalal M (2017) A worldwide SPT-based soil liquefaction triggering analysis utilizing gene expression programming and Bayesian proba-

bilistic method. J Rock Mech Geotech Eng 9(4):683–693. https://doi.org/10.1016/j.jrmge.2017.03.011
93. Hajihassani M, Abdullah SS, Asteris PG, Armaghani DJ (2019) A gene expression programming model for predicting tunnel convergence. Appl Sci 9(21):4650. https://doi.org/10.3390/app9214650
94. Kashani AR, Chiong R, Mirjalili S, Gandomi AH (2021) Particle swarm optimization variants for solving geotechnical problems: review and comparative analysis. Arch Comput Methods Eng 28(3):1871–1927. https://doi.org/10.1007/s11831-020-09442-0
95. Chao Z et al (2023) Artificial intelligence algorithms for predicting peak shear strength of clayey soil-geomembrane interfaces and experimental validation. Geotext Geomembr 51(1):179–198. https://doi.org/10.1016/j.geotexmem.2022.10.007
96. Tien Bui D, Hoang N-D, Nhu V-H (2019) A swarm intelligence-based machine learning approach for predicting soil shear strength for road construction: a case study at Trung Luong National Expressway Project (Vietnam). Eng Comput 35(3):955–965. https://doi.org/10.1007/s00366-018-0643-1
97. Obianyo II et al (2020) Multivariate regression models for predicting the compressive strength of bone ash stabilized lateritic soil for sustainable building. Constr Build Mater 263:120677. https://doi.org/10.1016/j.conbuildmat.2020.120677
98. Qu P-F, Zhang L-M, Zhu Q-Z (2023) Meta-modeling of fractional constitutive relationships for rocks based on physics-induced machine learning. Int J Numer Anal Methods Geomech 47(6):1000–1021. https://doi.org/10.1002/nag.3502
99. Zhang P, Yin Z-Y, Sheil B (2023) Interpretable data-driven constitutive modeling of soils with sparse data. Comput Geotech 160:105511. https://doi.org/10.1016/j.compgeo.2023.105511
100. Atef A, Hossain Z (2025) Assessing the impact of rice husk ash on soil strength in subgrade layers: a novel approach to sustainable ground engineering. Sustainability 17(12):5457. https://doi.org/10.3390/su17125457
101. Chen R, Congress SSC, Cai G, Duan W, Liu S (2021) Sustainable utilization of biomass waste-rice husk ash as a new solidified material of soil in geotechnical engineering: a review. Constr Build Mater 292:123219. https://doi.org/10.1016/j.conbuildmat.2021.123219
102. Mendes BC et al (2021) Application of eco-friendly alternative activators in alkali-activated materials: a review. J Build Eng 35:102010. https://doi.org/10.1016/j.jobe.2020.102010
103. Baghbani A, Soltani A, Kiany K, Daghistani F (2023) Predicting the strength performance of hydrated-lime activated rice-husk-ash-treated soil using two grey-box machine learning models. Geotechnics 3(3):894–920. https://doi.org/10.3390/geotechnics3030048
104. Pintelas E, Livieris IE, Pintelas P (2020) A grey-box ensemble model exploiting black-box accuracy and white-box intrinsic interpretability. Algorithms 13(1):17. https://doi.org/10.3390/a13010017
105. Esmaeili-Falak M, Katebi H, Vadiati M, Adamowski J (2019) Predicting triaxial compressive strength and Young's modulus of frozen sand using artificial intelligence methods. J Cold Reg Eng 33(3):04019007. https://doi.org/10.1061/(ASCE)CR.1943-5495.0000188
106. Lawal AI, Kwon S (2021) Application of artificial intelligence to rock mechanics: an overview. J Rock Mech Geotech Eng 13(1):248–266. https://doi.org/10.1016/j.jrmge.2020.05.010
107. Khajehzadeh M, Taha MR, Keawsawasvong S, Mirzaei H, Jebeli M (2022) An effective artificial intelligence approach for slope stability evaluation. IEEE Access 10:5660–5671. https://doi.org/10.1109/ACCESS.2022.3141432
108. Yao Q, Tu Y, Yang J, Zhao M (2024) Hybrid XGB model for predicting unconfined compressive strength of solid waste-cement-stabilized cohesive soil. Constr Build Mater 449:138242. https://doi.org/10.1016/j.conbuildmat.2024.138242
109. Onyelowe KC, Moghal AAB, Ebid A, Rehman AU, Hanandeh S, Priyan V (2024) Estimating the strength of soil stabilized with cement and lime at optimal compaction using ensemble-based multiple machine learning. Sci Rep 14(1):15308. https://doi.org/10.1038/s41598-024-66295-4

110. Priyadarshee A, Chandra S, Gupta D, Kumar V (2020) Neural models for unconfined compressive strength of kaolin clay mixed with pond ash, rice husk ash and cement. J Soft Comput Civil Eng 4(2):85–102. https://doi.org/10.22115/scce.2020.223774.1189
111. Mohammed AMA et al (2025) Explainable artificial intelligence for predicting the compressive strength of soil and ground-granulated blast-furnace slag mixtures. Results Eng 25:103637. https://doi.org/10.1016/j.rineng.2024.103637
112. Aslan Topçuoğlu Y (2025) Prediction of the effect of fly ash on the unconfined compressive strength of basalt fiber reinforced clay using artificial neural networks. Processes 13(1):157. https://doi.org/10.3390/pr13010157
113. Shariatmadari N, Hasanzadehshooiili H, Ghadir P, Saeidi F, Moharami F (2021) Compressive strength of sandy soils stabilized with alkali-activated volcanic ash and slag. J Mater Civil Eng 33(11):04021295. https://doi.org/10.1061/(ASCE)MT.1943-5533.0003845
114. Mozumder RA, Laskar AI (2015) Prediction of unconfined compressive strength of geopolymer stabilized clayey soil using artificial neural network. Comput Geotech 69:291–300. https://doi.org/10.1016/j.compgeo.2015.05.021
115. Eyo EU, Abbey SJ, Booth CA (2022) Strength predictive modelling of soils treated with calcium-based additives blended with eco-friendly pozzolans—a machine learning approach. Materials 15(13):4575. https://doi.org/10.3390/ma15134575
116. Tabarsa A, Latifi N, Osouli A, Bagheri Y (2021) Unconfined compressive strength prediction of soils stabilized using artificial neural networks and support vector machines. Front Struct Civil Eng 15(2):520–536. https://doi.org/10.1007/s11709-021-0689-9
117. Binal A, Binal BE (2020) Ternary diagrams for predicting strength of soil ameliorated with different types of fly ash. Arab J Sci Eng 45(10):8199–8217. https://doi.org/10.1007/s13369-020-04669-2
118. Javdanian H, Soltani N, Shams G, Ghorbani S (2021) Investigating the monotonic behavior of fiber-reinforced soil under triaxial compression using experimental modeling. Model Earth Syst Environ 7(2):943–952. https://doi.org/10.1007/s40808-020-00920-9

Chapter 7
Modeling Pressuremeter Modulus with Genetic Expression Programming: A Predictive Framework

Mohammad Jafar Hasanpour, Jafar Bolouri-Bazaz, S. Danial Mohammadzadeh, and Nader Karballaeezadeh

Abstract The pressuremeter test is one of the most reliable in situ methods in geotechnical engineering. It provides key parameters for evaluating bearing capacity and settlement behavior of foundations. This study proposes a gene expression programming (GEP) model to predict the pressuremeter modulus (E_p) based on three easily measurable soil properties, including depth from the ground surface, natural moisture content, and soil unit weight. The developed model achieved strong predictive performance: $R^2 = 0.9672$, normalized RMSE (NRMSE) = 0.0911, and performance index (PI) = 0.0459 for the testing dataset. These measures confirm the high level of reliability of the model. To better understand the role of input variables, sensitivity analyses using both Sobol and SHAP techniques were performed. These analyses consistently identified depth as the most influential factor, followed by moisture content and soil unit weight. A comparative evaluation with similar works in the literature further demonstrated that the proposed GEP model strikes an effective balance between accuracy, dataset adequacy, and model simplicity. These results highlight the model's potential as a practical and reliable tool for preliminary geotechnical design.

M. J. Hasanpour · J. Bolouri-Bazaz
Department of Civil Engineering, Ferdowsi University of Mashhad, Mashhad, Iran
e-mail: Bolouri@um.ac.ir

S. Danial Mohammadzadeh
Department of Civil Engineering, Ferdowsi University of Mashhad, Mashhad, Iran

Department of Civil Engineering, Architecture and Urbanism, Sadjad University of Technology, Mashhad, Iran

Department of Civil Engineering, Faculty of Engineering, Eqbal Lahoori Institute of Higher Education, Mashhad, Iran
e-mail: Danial.Mohammadzadehshadmehri@mail.um.ac.ir

N. Karballaeezadeh (✉)
Faculty of Civil Engineering, Shahrood University of Technology, Shahrood, Iran
e-mail: N.karballaeezadeh@shahroodut.ac.ir

M. Mirrashid et al. (eds.), *Intelligent Predictive Systems*, Emerging Trends in Mechatronics, https://doi.org/10.1007/978-981-95-7766-8_7

Keywords Geotechnical engineering · Pressuremeter modulus · Predictive modeling · Gene expression programming (GEP) · Sobol sensitivity analysis · SHapley Additive exPlanations (SHAP)

7.1 Introduction

The two primary methods employed in geotechnical engineering to assess soil deformation modulus are field and laboratory testing. Engineers can calculate one of the crucial characteristics of soils, named modulus of deformation, using both methods. In situ tests, however, are more reliable [1–4]. During in situ tests, the soil is less disturbed, and therefore, the natural condition of the ground is preserved. The pressuremeter test (PMT) is unique among in situ tests as it applies a horizontal load to the soil, and it is considered one of the most reliable methods because of its defined boundary conditions. The pressuremeter device introduced in 1957 by Menard has three main parts, including one control unit, one probe, and one tubing (which connects the control unit and the probe) [5]. The principle of Menard's pressuremeter involves the radial expansion of a cylindrical probe positioned in a prepared borehole. By subjecting the borehole wall to uniform pressure and monitoring its displacements, a stress–strain curve is generated. Then the pressuremeter modulus is determined from this curve. More details can be found in ASTM D4719-20 [6]. Test results are very valuable and reliable because the way in which the load is applied to the soil is very similar to loading in nature.

Pressuremeter tests are in situ strain–stress tests based on the measurement of radial volumes in the borehole and take place inside the borehole. There are two ways to conduct this test: (1) Stress-controlled method with the same pressure step increasing, and (2) Strain-controlled method with the same pressure volume increasing. In a pressuremeter test, a cylindrical chamber is placed inside a pre-drilled borehole. While increasing the volume of the chamber, the volume changes are measured against the pressure. The chamber (probe) expands by increasing the same steps of pressure (first method) or by increasing the same steps of volume (second method). The test continues until the soil reaches the point of yielding and large deformations [7]. The final limit of pressure (P_L) is predicted from the last readings, and the coefficient of pressuremeter (E_p) will be calculated from the reading of volume-pressure changes during the test. The analysis of the obtained results is utilized in the bearing capacity calculation, foundation settlement, and behavior of soil under horizontal load [5, 8, 9]. The pressuremeter modulus is determined by the following equation [10]:

$$E_p = 2 \times (1 + \upsilon) \times (V_m + V_0) \times \frac{\Delta V}{\Delta P} \tag{7.1}$$

where E_p (kPa) is the modulus of the pressuremeter. υ represents the Ration of Poisson. V_m and V_0 are the corrected volume reading in the center of the ΔV volume increase and the volume of the uninflated probe at the ground surface, respectively.

ΔP represents the corrected pressure increment in the central section of the straight portion of the pressure–volume curve. ΔV denotes the corrected volume increment in the same section and corresponds to the ΔP increase.

Although practical methods have their inherent accuracy, they suffer from some difficulties, including being time-consuming and costly, requiring specialist technicians, etc. Therefore, developing indirect approaches has gained increasing attention from researchers while determining the E_p. As of now, numerous attempts have been made to find relations between E_p and soil characteristics. Empirical formulation is always considered the first answer for such engineering problems. Table 7.1 provides the results of many studies aimed at creating an empirical equation for estimating E_p. This table confirms the existence of meaningful relationships between soil properties, including results from other in site tests such as cone penetration test (CPT), the standard penetration test (SPT), and modulus derived from the pressuremeter tests. In the mentioned studies in Table 7.1, E_p has been considered as a function of different parameters like cone resistance (q_c), standard penetration number (N_{60}), moisture content (ω), plasticity index (PI), depth of test (D), gravel-grained content (G), sand-grained content (S), and atmospheric pressure (P_a).

Empirical equations are not able to capture the highly nonlinear interactions between E_p and input parameters, i.e., the soil parameters; therefore, more sophisticated methods are demanded to consider this behavioral complexity. Artificial intelligence (AI) techniques are appropriate solutions for this need. AI has emerged as a powerful field that can develop computational techniques, inspired by natural processes, in order to solve complex real-world problems in civil engineering [19–24]. In recent years, geotechnical researchers have employed numerous

Table 7.1 Summary of empirical models for E_p prediction

Equation/model	Soil type	Reference	Year
$E_p = 0.442q_c + 2.221$ $E_p = 371.1 - 1.286N_{60}$	Sandy silt and sandy gravel	[11]	2023
$E_p = 1.673G - 3.21\omega + 1.287D + 67.083$ $E_p = 0.223S - 3.006\omega + 1.586G + 1.218D + 60.848$	Clay	[12]	2018
$E_p = 37.83 + 29.5 \ln (N_{60}) - 0.0086\omega - 0.0034\omega^2 + 8.55(\ln(N_{60}))^2 - 0.018\omega \ln (N_{60})$	Clay	[13]	2017
$E_p = P_a(9.8N_{60} - 94.3)$	Silty sand	[14]	2014
$E_p = P_a(10N_{60} - 26.7)$	Silty clay		
$E_p = 0.2885N_{60}^{1.4}$	Clay	[15]	2012
$E_p = 1.24N_{60}^{0.94} - 11.04 \ln \omega + 37.9$			
$E_p = 0.68\text{PI} + 0.014(N_{60})^{2.067} - 10.44 \ln (\omega) + 23.82$			
$E_p = 1.61N_{60}^{0.71}$	Clay	[16]	2010
$E_p = 1.33N_{60}^{0.77}$	Sand		
$E_p = 388.67N_{60} + 4554$	Sandy silty clay	[17]	2008
$E_p = 2.5q_c$	Clay	[18]	1985
$E_p = 1.15q_c$	Sand		

computer-aided models with the hope of enhancing the existing procedures and exploring alternatives for simulating the behavior of soils [25–29], specifically for modeling E_p [12, 13, 30–32]. Among the various branches of AI, evolutionary computation has gained widespread recognition, particularly Genetic Programming (GP), which Koza introduced in the early 1990s [33]. GP employs evolutionary regulations where the solutions are computer programs rather than binary strings that make it effective for modeling nonlinear systems. However, traditional GP faces challenges such as code bloat and difficulties in genetic operations [34]. To address these issues, Ferreira (2001) proposed Genetic Expression Programming (GEP), a variant of GP that encodes solutions as linear chromosomes which are later expressed as expression trees. This representation enhances efficiency, flexibility, and robustness, and makes GEP a widely adopted method for predictive modeling and optimization in engineering and scientific domains [35, 36]. The primary focus of this study is the application of GEP for predicting the E_p. Only a limited number of studies have addressed this task using GEP, as summarized in Table 7.2. Rashed et al. [10] developed an equation with four input parameters, including fine-grained content, coefficient of uniformity, moisture content, and soil dry unit weight, based on 106 pressuremeter test results. Their model achieved an acceptable coefficient of determination ($R^2 = 0.811$), but exhibited relatively high error values because MAE and RMSE were 49.58 and 72.14, respectively. In the same year, Alavi et al. [37] utilized 114 test results to construct a more complex equation, requiring 10 input parameters. Despite its complexity, their model demonstrated higher accuracy with an R^2 of 0.941, and with lower error levels (MAE = 5.48, RMSE = 6.72). More recently, Razavi et al. [12] employed 62 clay soil test results to propose a model relating E_p to test depth, coarse (gravel and sand) and fine-grained content, and moisture content. Their approach yielded superior performance, including $R^2 = 0.981$, MAE = 6.915, and RMSE = 7.917.

The review of the literature shows that there is limited attention to employing GEP as a predictive tool for estimating the E_p. As a consequence, the authors aim to bridge this gap by exploring further with regard to the GEP's potential in this field. The present study introduces a novel GEP-based model for E_p prediction. Unlike previous approaches, the proposed model not only attains a high, acceptable level of predictive accuracy but also relies on fewer input variables, making it both simpler and more practical for broader geotechnical applications.

The structure of this paper is as follows: Sect. 7.2 focuses on the data preparation and explains the method employed by the authors. Additionally, all measures used to evaluate the performance of the developed model are introduced. Next, in Sect. 7.3, the results of the model are presented. In this part of the paper, a sensitivity analysis is added to compare the impact of input variables on the model prediction. Finally, the conclusion is the final section of this paper.

Table 7.2 Summary of GEP-based models for pressuremeter modulus prediction

Equation/model	Soil type	Reference	Year
$E_{\mathrm{p}} = 28 - 4\mathrm{FC} + 4\gamma_{\mathrm{d}}\left(\dfrac{\dfrac{288}{\omega \times \gamma_{\mathrm{d}}^{3}}\left(\dfrac{12\sqrt{C_{\mathrm{u}}} + 12\mathrm{FC}}{\gamma_{\mathrm{d}}} + 7\right)^{2} + 60}{\mathrm{FC}}\right)$	All soil types	[10]	2012
$E_{\mathrm{p}} = \dfrac{\mathrm{CC}}{\mathrm{FC}}\left(2 + \left(\mathrm{LL} + D_{10}\left(D_{60} - \dfrac{\mathrm{CC}}{\mathrm{FC}}e^{\left(\sqrt[3]{\mathrm{LL}+8} - C_{\mathrm{C}}\right)}\right)^{2}\right)\right) + \left(\dfrac{4D_{60}^{2}}{20.085\left(\gamma_{\mathrm{d}} - \mathrm{LL}\right) + \omega e^{\left(D_{60} - D_{10}\right)}} - 4\right)^{2}$ $+ D_{10}\omega\left(2C_{\mathrm{C}} - 3\dfrac{\mathrm{CC}}{\mathrm{FC}}\left(C_{\mathrm{C}} - \mathrm{LL} + \gamma_{\mathrm{d}}\right)\right) + \left(\dfrac{\mathrm{CC}}{\mathrm{FC}}\right)^{3} D_{10}\left(\gamma_{\mathrm{d}} - D_{60} + \sqrt[3]{\mathrm{LL}}\left(\dfrac{\mathrm{CC}}{\mathrm{FC}} + 0.25\omega - \gamma_{\mathrm{d}}\right)\right) - \dfrac{\left(\dfrac{\mathrm{CC}}{\mathrm{FC}} + C_{\mathrm{C}}\right)\left(C_{\mathrm{C}} + \omega\right)}{\gamma_{\mathrm{d}} - \left(C_{\mathrm{C}} - C_{\mathrm{u}}\dfrac{\mathrm{CC}}{\mathrm{FC}}\right)\left(5D_{60} - \omega\right)} - D_{60}$	All soil types	[37]	2012
$E_{\mathrm{p}} = 26.988 - \sqrt[3]{G^{2}\left(16.698 - F\right)} + 0.529\left(1087 + S - 2\omega\right) \times \left(\dfrac{D}{\omega}\right) + G + \sqrt[3]{8.046 - D}$	Clay	[12]	2018

7.2 Method and Material

7.2.1 Data Acquisition

In this study, first, the most influential factors on the behavior of the soil deformation modulus were identified through a comprehensive literature review [38–42], including intrinsic soil properties such as FC, CC, D_{10}, D_{30}, D_{60}, C_u, C_c, and LL, and state-related parameters such as ω and γ. Involving such specifications helps enhance the reliability of estimating the pressuremeter modulus with an acceptable level of error. Next, the final factors were selected for establishing the E_p estimation modeling:

$$E_p\left(\text{kg}/\text{cm}\right)=f\left(H,\omega,\gamma\right) \tag{7.2}$$

where H, ω, and γ stand for depth (m), moisture content (%), and soil unit weight (g/cm^3), respectively.

This combination of input factors is rooted in the fact that the authors aimed to develop a simple and user-friendly model. In addition to reducing the number of required tests, such modeling can lower project costs, save time, and simultaneously deliver continuous information about subsurface conditions. Therefore, it assists designers in enhancing their knowledge base.

To collect the data for developing the E_p model, 103 PMT tests were conducted on different sites (with various soil types) in Mashhad, Khorasan province, Iran. All tests were performed using the Menard pressuremeter device under standard conditions (ASTM D4719) and based on the stress-controlled method. Before the testing phase, the pressuremeter device was calibrated to compensate for pressure loss (caused by the hardness of the probe membrane) and volume loss (caused by the expansion of the tubes and the compressibility of the device components, including the tube and its internal components). The diameter of the device's probe was 58 mm, and the boreholes were drilled at the diameters of 110, 115, and 125 mm. Additionally, required laboratory tests have been performed to determine the physical and natural characteristics of the soils.

7.2.2 Statistical Analysis of the Gathered Dataset

The statistical characteristics of the data used are according to Table 7.3. This table summarizes the descriptive statistics of the dataset, including central tendency, dispersion, and distribution shape parameters for all involved variables. The mean values indicate that the soils investigated are generally located at intermediate depths (19.33 m), with moderate moisture content (9.25%), relatively consistent unit weight (1.79 g/cm^3), and an average pressuremeter modulus of 613.74 kg/cm^2. The standard deviations reveal high variability in H and E_p, suggesting substantial heterogeneity across sampling points, while γ remains highly uniform. The minimum

Table 7.3 Descriptive statistics analysis

Parameter	H	ω	γ	E_p
Mean	19.33	9.25	1.79	613.74
Standard deviation	10.41	6.11	0.074	272.70
Kurtosis	−1.24	1.04	−0.35	−1.005
Skewness	0.13	1.14	0.126	0.212
Range	34	27.7	0.37	1068
Minimum	5	1.3	1.61	198
Maximum	39	29	1.98	1266

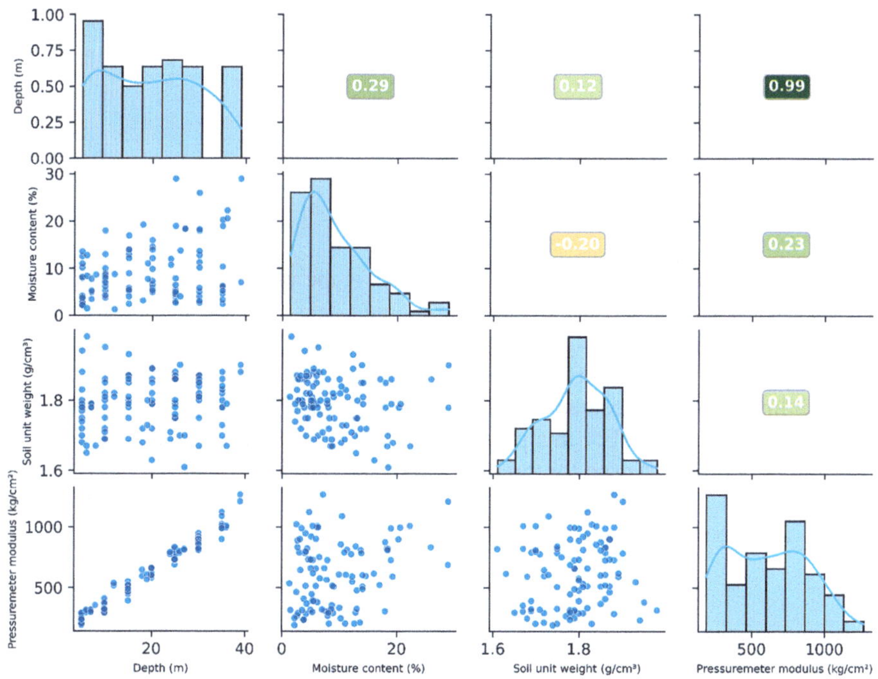

Fig. 7.1 Correlation matrix of variables

and maximum values further highlight this variability, with H ranging from 5 to 39 m and E_p ranging from 198 to 1266 kg/cm^2. Skewness and kurtosis values indicate that most variables are approximately normally distributed, with slight positive skewness observed for ω and nearly symmetrical distributions for H, γ, and E_p. Negative kurtosis for H and E_p indicates relatively flat distributions, while ω exhibits a more peaked distribution. Collectively, these results show that H and E_p are the most variable parameters in the dataset, whereas γ is highly consistent, providing important context for interpreting subsequent correlation and modeling analyses.

Figure 7.1 presents a comprehensive correlation matrix illustrating the statistical relationships among all variables. The diagonal panels display the distribution of each variable with overlaid kernel density curves, highlighting the spread and central tendency of the data. H and E_p exhibit wide distributions, whereas γ shows a narrow range, reflecting its uniformity across the dataset. The lower triangular panels provide pairwise scatter plots, where a strong linear relationship is clearly observed between H and E_p, while other variable pairs reveal weaker or more dispersed associations. The upper triangular panels report Pearson correlation coefficients, color-coded according to strength and direction. A nearly perfect positive correlation (0.99) is identified between H and E_p, confirming that pressuremeter modulus increases with depth. A moderate positive correlation is observed between ω and E_p, while a lower positive correlation exists between γ and E_p. Overall, the figure underscores the dominant influence of H on E_p while indicating that moisture content contributes minimally to variations in E_p. The combined use of histograms, scatter plots, and annotated correlation coefficients provides a holistic representation of the dataset, facilitating both visual and quantitative interpretation of the underlying relationships.

7.2.3 Data Splitting

One of the important concepts in AI techniques is the phenomenon of overfitting. This means that good performance on the training data accompanied by weak performance on the test data reveals that the model has overfit the training set. To prevent the phenomenon of overfitting, the data should be split into two categories: test and training. Then, the accuracy of the presented model should be checked on these two categories [43, 44]. For this purpose, among the 103 available data points, 72 data points (70%) were used for training, and 31 data points (30%) were used for the final testing of the models. This classification has also been made based on the reviews and suggestions of researchers [45, 46].

7.2.4 GEP Model Development

GEP is an evolutionary computation method that draws inspiration from natural selection and biological evolution. Evolutionary algorithms such as Genetic Algorithms (GA) and Genetic Programming (GP) have been widely used in engineering and scientific fields to solve complex optimization and modeling problems. While GA operates on fixed-length strings and GP manipulates variable-length tree structures, GEP integrates the advantages of both by employing linear chromosomes that are expressed as expression trees during evaluation. This dual representation allows GEP to maintain computational efficiency while preserving the structural flexibility necessary to capture nonlinear relationships among variables. Owing to its adaptability and robustness, GEP has become a promising tool for modeling highly nonlinear systems in geotechnical and civil engineering research [47–51].

In this study, a GEP model was developed to predict the E_p from the input parameters of depth, moisture content, and soil unit weight. The modeling was implemented in Python within the Jupyter Notebook environment using the gplearn library, which provides symbolic regression capabilities through evolutionary computation. A schematic representation of the GEP modeling process is provided in Fig. 7.2. The figure illustrates the main stages of GEP, beginning with the

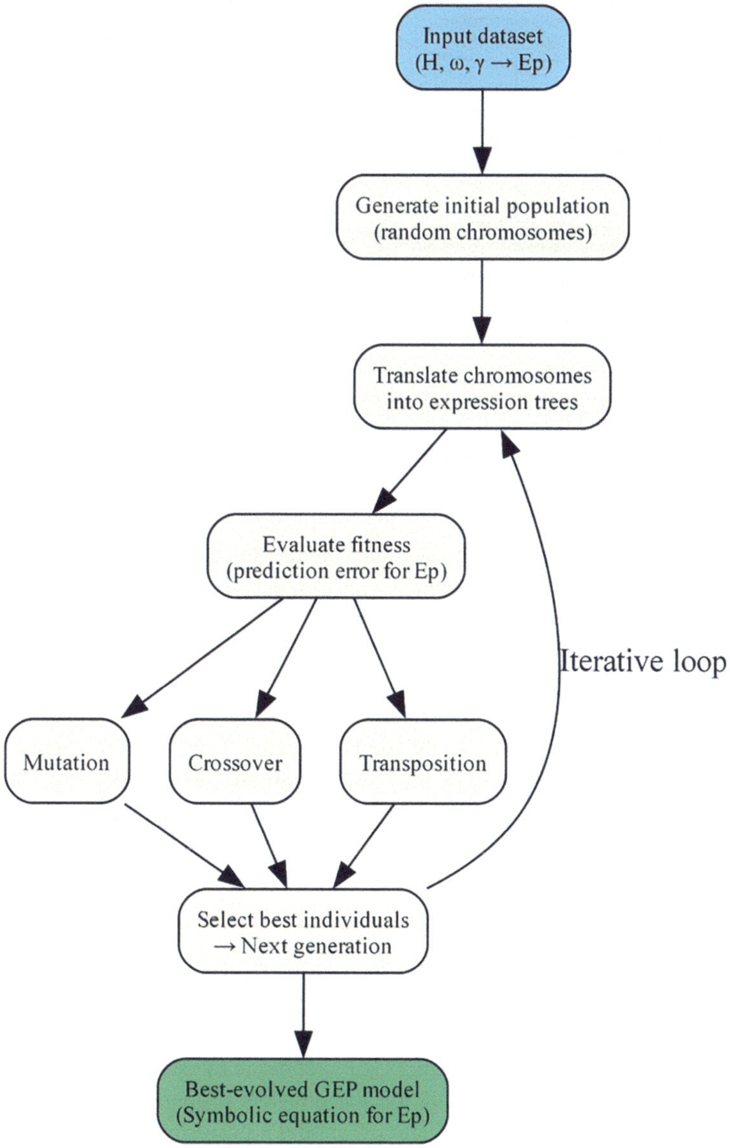

Fig. 7.2 Schematic representation of the GEP modeling process

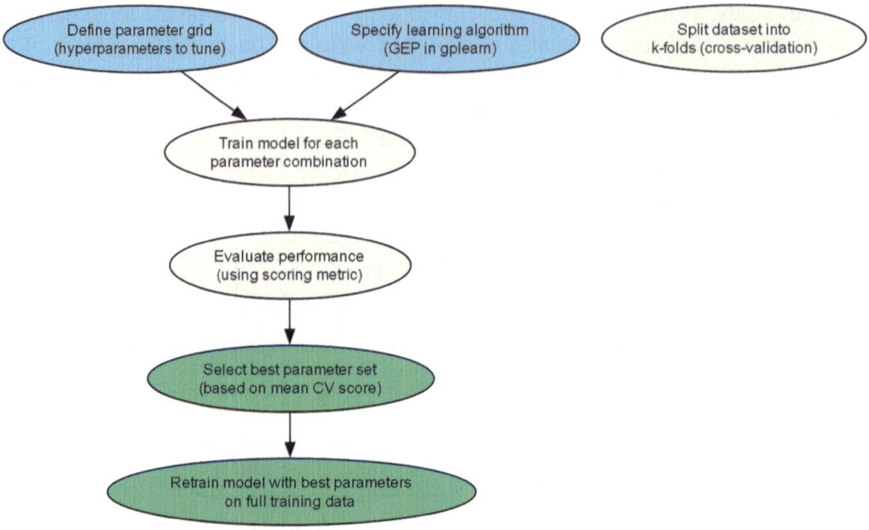

Fig. 7.3 Flowchart of the GridSearchCV procedure

initialization of a random population of chromosomes, followed by genetic operations such as mutation, crossover, and transposition, which are applied iteratively to evolve the population. Each chromosome is expressed as an expression tree and evaluated using a predefined fitness function, with the best-performing individuals propagated to the next generation. The evolutionary process proceeds iteratively until a predefined stopping condition, such as reaching the maximum number of generations or achieving performance convergence, is satisfied. This schematic clarifies the fundamental logic of GEP and highlights its capability to discover symbolic relationships between input and output variables without assuming a predefined functional form.

To ensure the reliability and efficiency of the model, hyperparameter optimization was carried out using the GridSearchCV module from the scikit-learn framework. This approach (Fig. 7.3) allowed systematic exploration of candidate hyperparameter sets, such as population size, number of generations, and function set, ultimately identifying the best configuration for achieving robust model performance.

7.2.5 Model Assessment Criteria

The objective of this study is to use the GEP technique to establish a novel relationship for estimating the soil E_p based on the physical properties of the soil. Three evaluation metrics are used to evaluate the model's performance: the performance index (PI), normalized root mean square error (NRMSE), and correlation coefficient (R^2). R^2 falls in [0, 1] with a general role: Closer to 1, higher accuracy of the model's performance. The problem with the root mean square error (RMSE), which is the inability to interpret the RMSE's absolute magnitude, especially when variables have different scales, is addressed by NRMSE. Lower values indicate better predictive accuracy and dependability. By normalizing RMSE, NRMSE offers a standardized measure of error, enabling fair comparisons across various models [52]. The PI is also employed to assess model performance, as it accounts for variations in both correlation and error. Its values lie within the range of [0, +∞], with smaller values indicating greater precision [29].

$$R^2 = 1 - \frac{\sum_{i=1}^{n} (h_i - t_i)^2}{\sum_{i=1}^{n} (h_i - \bar{h}_i)^2} \tag{7.3}$$

$$\text{NRMSE} = \frac{\text{RMSE}}{\text{mean}(h_i)} \tag{7.4}$$

$$\text{PI} = \frac{\frac{1}{|\bar{h}_i|}\sqrt{\text{MSE}}}{R+1} \tag{7.5}$$

where

$$\text{RMSE} = \sqrt{\text{MSE}} \tag{7.6}$$

$$\text{MSE} = \frac{\sum_{i=1}^{n} (h_i - t_i)^2}{n} \tag{7.7}$$

where h_i and t_i present the real and estimated values for the ith output, respectively. $\bar{h}_i$ is the average of h_i, and the total number of samples is represented by n.

7.3 Results and Discussion

7.3.1 Modeling Results

The purpose of this study is to present an AI model for estimating E_p using a method called GEP that combines GA and GP. The performance of GEP is dependent on its hyperparameters, just like other ML techniques. Thus, in recent years, hyperparameter tuning has become more and more popular. Among several optimization tools that have been introduced recently, the authors selected GridSearchCV for this study. The tuning strategy was applied to the most influential hyperparameters of GEP. Table 7.4 demonstrates the candidate values, which came from what has been

Table 7.4 Defined space in GridSearchCV

Hyperparameter	Suggested values
population_size	[500[a], 1000, 2000]
generations	[10, 30, 50[a]]
tournament_size	[10, 30, 50[a]]
p_point_mutation	[0.05, 0.1[a], 0.2]
p_hoist_mutation	[0.01, 0.05[a], 0.1]
p_subtree_mutation	[0.1[a], 0.2, 0.3]
p_crossover	[0.6, 0.7[a], 0.8]

[a] The tuned values

Table 7.5 Finalized hyperparameters

Hyperparameter	Selected values
population_size	500
generations	50
tournament_size	50
p_point_mutation	0.1
p_hoist_mutation	0.05
p_subtree_mutation	0.1
p_crossover	0.7
parsimony_coefficient	0.01
stopping_criteria	0.01
Verbose	1
function_set	'add', 'sub', 'mul', 'div'

suggested in the literature [37, 45, 47, 53] and the authors' experience, for tuning the GEP model. The GridSearchCV technique explored the defined spaces and found the optimal values for population_size, generations, tournament_size, p_crossover, p_subtree_mutation, p_hoist_mutation, and p_point_mutation hyperparameters. The implementation of the GridSearchCV technique was conducted using the sklearn library in Python, while SymbolicRegressor from gplearn.genetic, r2_score, 3, −1, and 2 were assigned to estimator, scoring, cv, n_jobs, and verbose, respectively. Other hyperparameters of GEP were designated based on the authors' experience (Table 7.5), especially about the function_set, where four basic operations, including addition, subtraction, multiplication, and division, were opted for since the authors believed in the simplest formulation. Finally, Eq. (7.8) presents the formula constructed based on the modeling output.

$$E_{\mathrm{p}} = 22.111\left(H - 0.981\right) + \left(2H + A + \left(\frac{\omega}{0.011} + \left(\frac{0.138 \times H}{\omega} - Y\right)\right)\right) \tag{7.8}$$

where

$$A = \frac{\dfrac{\dfrac{0.040434H^2}{\omega} - H + Y}{\omega + 2Y - H}}{\dfrac{2}{0.534} + \left(\dfrac{0.293H^2}{-Y^2 + 0.707YH}\right)} \times \left(\frac{\dfrac{0.293H^2}{\omega}}{0.086Y - \dfrac{0.647}{H}} + \frac{H(Y-\omega)}{\omega(0.138H - Y)}\right) \tag{7.9}$$

7.3.2 Modeling Performance Evaluation

The evaluation of the proposed model, GEP, is an important step in any scientific work. Generally, performance assessment focuses on two main points: how accurate the model was and how much error the model had. To this end, many different methods have been developed over the years that can be categorized into two main groups: (1) Statistical, and (2) Graphical. The first group deals with numerical criteria representing the accuracy and error of models based on a series of mathematical operations and considering one number as the measure of judgment, including measures introduced in Sect. 7.2.5. The second group handles visual discussion on the model performance. Scatter and error distribution plots are the most famous and common examples of this group.

In this study, the first step involved statistical analysis of the model prediction. Table 7.6 presents the values of the criteria introduced earlier as the measures of GEP output goodness. R^2 shows a great accuracy on the train dataset. Besides, NRMSE and PI reflect having a low error when the model is trained on the training dataset. Although all three criteria experienced a little drop in the test dataset, they still confirm the good performance of GEP in this study.

Regarding graphical assessment, the authors firstly employed a scatter plot (Fig. 7.4) to compare the E_p values predicted by GEP and the E_p values existing in the dataset. Figure 7.4 includes two plots for both the train and test datasets. Each plot shows all data of the original dataset versus their corresponding prediction, produced by GEP, as black points. For these points, a blue line was fitted (regression line). In fact, the regression line is an equivalent of scatter points and aims to ease the GEP performance judgment. This judgment would be possible by means of another line: a black dashed line, namely the 1:1 line. The 1:1 line is a reference line for the regression line and helps the authors decide whether GEP performed well or not. Closer to the 1:1 line, better performance of GEP. The 1:1 line has the equation of $Y = 1.0x + 0.0$; then the slope and intercept of the regression line must be near 1.0 and 0.0, respectively. For both plots in Fig. 7.4, it is vivid that GEP performed well.

Table 7.6 Model assessment measure values

	R^2	NRMSE	PI
Train dataset	0.9776	0.0619	0.0311
Test dataset	0.9672	0.0911	0.0459

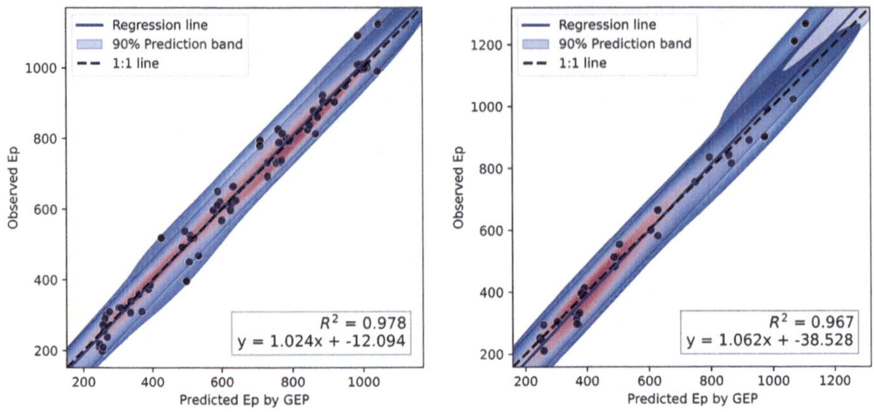

Fig. 7.4 Density scatter plot for regression analysis of the GEP model: the upper plot, training, and the lower plot, test

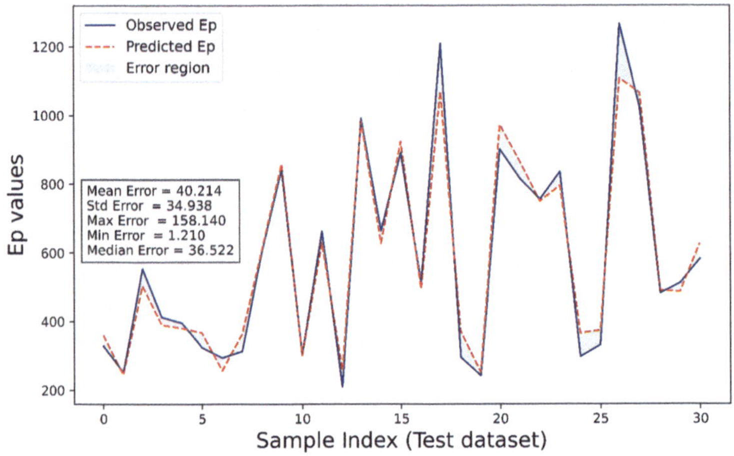

Fig. 7.5 Error analysis curve

Figure 7.5 illustrates the error analysis for the test dataset, which includes 31 samples. The observed and predicted values of E_p are plotted against the minimal gap between the estimated E_p values, red dashed line, and the measured E_p values, blue solid line, show that the GEP model provides accurate predictions for most test cases. The statistical error metrics confirm that errors are generally low and concentrated around the mean as they reported a mean absolute error of 40.214, a standard deviation of 34.938, and a median error of 36.522. Although the maximum error reached up to 158.140 for certain samples, there was a minimum possible error of 1.210, which indicates that the model is performing well in retrieving the general trend of E_p with reasonable errors at certain points.

7.3.3 Sensitivity Analysis

Sensitivity analysis aids in grasping how changes in independent (input) variable values affect the dependent (output) variable. Also, this analysis suits situations with multiple input variables, as in this study, where the relative influence of inputs is checked. In this study, sensitivity analysis checks how H, ω, and γ influence the E_p values predicted by the GEP model. In addition, the order of influence strength is another important finding that the analysis follows. Two popular methods were selected to implement the sensitivity analysis: Sobol and SHAP.

The Sobol method, a variance-based approach, is a sensitivity analysis technique used to assess the degree to which individual input parameters influence the variability of a model's output [54]. In this study, the Sobol technique results are expressed as two factors of the first-order sensitivity index (S_1) and the total-order sensitivity index (ST). Table 7.5. Finalized hyperparameters shows how much one input parameter impacts output variance. ST includes S_1, but also factors in how that input parameter interacts with others. The results of the Sobol analysis are shown in Fig. 7.6. This figure confirms that H is the dominant input variable, with the highest first-order sensitivity index, S_1 = 0.222, and an exceptionally large total effect of 4.974. Therefore, H has a dominant role in explaining the variance of the model output. After that, ω shows a moderate total effect of 0.788, although its main effect is comparatively small: S_1 = 0.022. Finally, γ demonstrates negligible sensitivity indices because S_1 and ST values are 0.002 and 0.014, respectively, highlighting its limited contribution. Additionally, the difference between S_1 and ST for all input variables reflects the influence of each input mainly arises through interactions with other inputs.

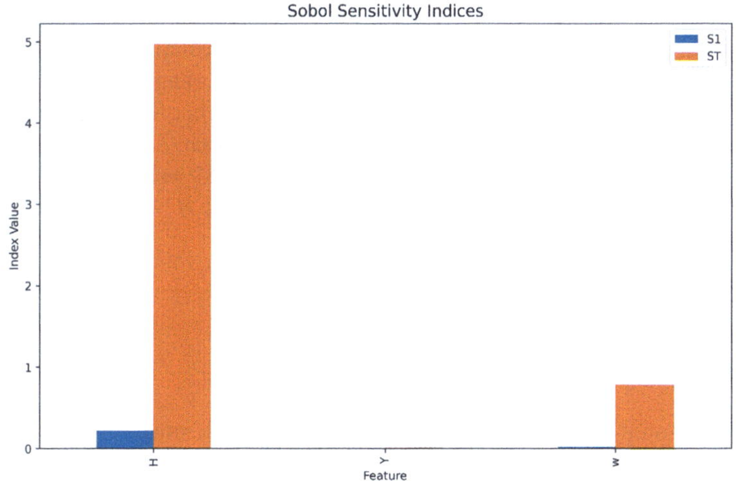

Fig. 7.6 Sobol sensitivity analysis results

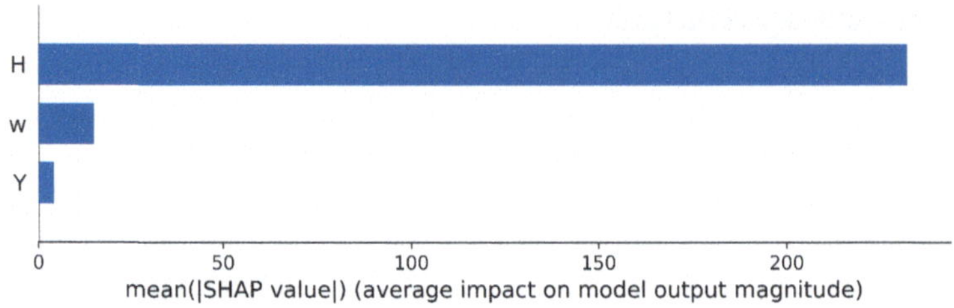

Fig. 7.7 Mean absolute SHAP values

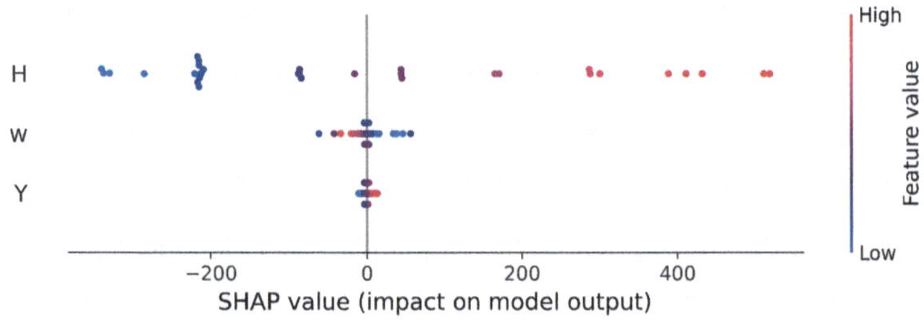

Fig. 7.8 SHAP beeswarm plot

The SHAP-based sensitivity analysis provides further evidence of input importance [25]. SHAP is a method that explains how individual predictions are made by the GEP model. Figure 7.7 supports that *H* ranks first, with a mean absolute SHAP value of 232.19. This suggests *H* has the strongest impact on the predicted E_p values across the dataset. In comparison, ω and γ with a mean absolute SHAP value of 15.18 and 4.22, respectively, play considerably smaller roles, though they still contribute marginally to the predictions. Figure 7.8, a beeswarm SHAP plot, shows which features have the biggest impact on the GEP's predictions. In addition, this plot helps the authors understand how the predicted value of an observed E_p changes as the value of a single input feature varies. *H* has the higher SHAP values, leading to a higher impact of *H* on the GEP predictions. After that, ω ranks as the second most influential input. Finally, γ stands with lower SHAP values. As a result, the SHAP analysis not only confirms the critical influence of *H* but also highlights the relatively minor effects of ω and γ in the GEP model.

7.3.4 Comparative Analysis

Comparative analysis forms a key part of scientific work since it situates the presented model in relation to current published work. By comparing results with prior research, one can better show the strength, efficiency, and real-world use of the new model. In this regard, the proposed GEP model was compared with similar works

Table 7.7 Comparison of the performance of the proposed GEP model with related studies in the literature

References	Method	R^2	Dataset size	Number of inputs
Alavi et al. [37]	GEP	0.965	114	10
Rashed et al. [10]	LGP[a]	0.908	106	4
Razavi et al. [12]	GEP	0.981	62	5
This study	*GEP*	*0.9672*	*103*	*3*

[a] *LGP* linear genetic programming

available in the literature. The detailed comparison of model performance, dataset size, and input requirements is summarized in Table 7.7. Alavi et al. [37] achieved a high predictive accuracy ($R^2 = 0.965$) using 114 data samples; however, their model relied on 10 input variables, which may complicate data collection and reduce the model's practicality. Rashed et al. [10], who employed linear genetic programming (LGP), reported a lower accuracy ($R^2 = 0.908$) despite using a comparable dataset size and fewer inputs (four), suggesting that reducing the number of variables without maintaining sufficient model complexity can compromise predictive power. The model developed by Razavi et al. [12] obtained the highest reported accuracy: $R^2 = 0.981$. However, their model relied on a relatively small dataset (62 samples) and five inputs, raising concerns about the robustness and generalizability of their findings. Compared to the discussed studies, the proposed GEP model in this study achieved an R^2 of 0.9672 with a reasonably large dataset of 103 samples while requiring only three input variables, the lowest among all reviewed works. This balance among accuracy, dataset adequacy, and parsimony in input requirements underscores both the efficiency and practical utility of the proposed GEP model, confirming its competitiveness and acceptability for predictive applications.

7.4 Conclusion

This study presented a GEP model for predicting the pressuremeter modulus (E_p) using three soil parameters: depth from the ground surface, natural moisture content, and soil unit weight. The developed model demonstrated high predictive accuracy, with performance measures of $R^2 = 0.9672$, NRMSE = 0.0911, and PI = 0.0459 on the testing dataset, confirming its reliability for practical applications. Sensitivity analyses conducted through Sobol and SHAP methods consistently identified depth as the most influential input variable, followed by moisture content, while soil unit weight had the least impact. A comparative study with similar works in the literature further revealed that the proposed model offers a favorable balance between prediction accuracy, dataset adequacy, and input simplicity, establishing its competitiveness and efficiency. Overall, the findings confirm that the developed GEP model can serve as a robust and practical tool for preliminary geotechnical design and decision-making, reducing the need for extensive in situ testing while maintaining a reliable level of confidence.

For future research, extending the dataset to cover wider soil types and geological conditions would enhance the generalizability of the model. In addition, exploring hybrid approaches that integrate GEP with other machine learning techniques may further improve predictive accuracy. Finally, validating the model through large-scale field applications could strengthen its reliability and promote its adoption in routine geotechnical practice.

Conflict of Interest The authors declare that they have no conflicts of interest related to this research.

Data Availability Statement The data that support the findings of this study are available from the corresponding author upon reasonable request.

Funding Statement The authors received no financial support for the research, authorship, and/or publication of this article.

AI Use Disclosure Statement During the preparation of this work, the authors used Grammarly for proofreading aims, including grammar and spelling checks. After using this tool, the authors reviewed and edited the content as needed and took full responsibility for the content of the published article.

References

1. Reznik YM (1995) Comparison of results of oedometer and plate load tests performed on collapsible soils. In: Genesis and properties of collapsible soils. Springer, pp 383–408
2. Murthy V (2002) Geotechnical engineering: principles and practices of soil mechanics and foundation engineering. CRC Press
3. Mohammadzadeh DS et al (2025) Analysis and modeling of soil retaining structure for urban train stations. In: Advancements in underground infrastructures. CRC Press, pp 147–183
4. Maroof A et al (2024) Investigation of empirical and analytical methods accuracy for surface settlement prediction in train tunnel excavation projects. Acta Polytech Hung 21(1):167
5. Menard LF (1957) An apparatus for measuring the strength of soils in place. PhD thesis, University of Illinois
6. ASTM D4719-20 (2020) Standard test methods for prebored pressuremeter testing in soils. ASTM International, West Conshohocken, PA
7. ASTM D4719-07 (2000) Standard test method for prebored pressuremeter testing in soils. American Society for Testing & Materials Standard
8. Baguelin F, Jezeguel JF, Shields DH (1978) The pressuremeter and foundation engineering. Trans Tech Publications
9. Briaud J-L (2007) Spread footings in sand: load settlement curve approach. J Geotech Geoenviron 133(8):905–920
10. Rashed A, Bazaz JB, Alavi AH (2012) Nonlinear modeling of soil deformation modulus through LGP-based interpretation of pressuremeter test results. Eng Appl Artif Intell 25(7):1437–1449
11. Moghadari Poor M, Azarafza M, Derakhshani R (2023) A correlation based on pressuremeter, SPT and CPT tests for characterizing of coastal alluvium: a study for phase 14 South Pars, Iran. MethodsX 10:101938

12. Razavi S et al (2018) Proposing new relationships to estimate the pressuremeter modulus of cohesive and cohesionless media. Innov Infrastruct Solut 3(1):67
13. Ziaie Moayed R, Kordnaeij A, Mola-Abasi H (2018) Pressuremeter modulus and limit pressure of clayey soils using GMDH-type neural network and genetic algorithms. Geotech Geol Eng 36(1):165–178
14. Cheshomi A, Ghodrati M (2015) Estimating Menard pressuremeter modulus and limit pressure from SPT in silty sand and silty clay soils. A case study in Mashhad, Iran. Geomech Geoeng 10(3):194–202
15. Kayabasi A (2012) Prediction of pressuremeter modulus and limit pressure of clayey soils by simple and non-linear multiple regression techniques: a case study from Mersin, Turkey. Environ Earth Sci 66(8):2171–2183
16. Bozbey I, Togrol E (2010) Correlation of standard penetration test and pressuremeter data: a case study from Istanbul, Turkey. Bull Eng Geol Environ 69(4):505–515
17. Yagiz S, Akyol E, Sen G (2008) Relationship between the standard penetration test and the pressuremeter test on sandy silty clays: a case study from Denizli. Bull Eng Geol Environ 67(3):405–410
18. Briaud J et al (1985) Correlation between pressuremeter data and other parameters. Civil Engineering Research Report. Texas A&M University, College Station, TX
19. Koopialipoor M, Tootoonchi H, Marto A, Faizi K, Armaghani DJ (2020) Various effective factors on peak uplift resistance of pipelines in sand: a comparative study. Int J Geotech Eng. 24(7):820–827
20. Kumar P, Samui P, Armaghani DJ, Roy SS (2024) Second-order reliability analysis of an energy pile with CPT data. J Build Eng 95:110165
21. Asteris PG, Armaghani DJ (2025) An empirical-driven machine learning (EDML) approach to predict PPV caused by quarry blasting. Bull Eng Geol Environ 84(4):200
22. He B, Li J, Armaghani DJ, Hashim H, He X, Pradhan B, Sheng D (2025) The deep continual learning framework for prediction of blast-induced overbreak in tunnel construction. Expert Syst Appl 264:125909
23. Yu Huat C, Armaghani DJ, Lai SH, Rasekh H, He X (2023) Simulation of surface settlement induced by parallel mechanised tunnelling. Sustainability 15(17):13265
24. Yari M, Khandelwal M, Abbasi P, Koutras EI, Armaghani DJ, Asteris PG (2024) Applications of soft computing methods in backbreak assessment in surface mines: a comprehensive review. Comput Model Eng Sci 140(3):2207–2238
25. Maroof A et al (2025) Prediction of the wetting-induced compression of collapsible soils using ensemble machine learning. Indian Geotech J:1–20. https://doi.org/10.1007/s40098-025-01335-y
26. Zhang W et al (2021) Application of deep learning algorithms in geotechnical engineering: a short critical review. Artif Intell Rev 54(8):5633–5673
27. Gao W (2024) The application of machine learning in geotechnical engineering. Appl Sci 14:4712
28. Harle SM, Wankhade RL (2025) Machine learning techniques for predictive modelling in geotechnical engineering: a succinct review. Discov Civil Eng 2(1):1–21
29. Jalali H et al (2024) Prediction of vertical displacement for a buried pipeline subjected to normal fault using a hybrid FEM-ANN approach. Front Struct Civ Eng 18(3):428–443
30. Wu M et al (2021) Prediction of limit pressure and pressuremeter modulus using artificial neural network analysis based on CPTU data. Arab J Geosci 14(1):2
31. Alidousti Shahraki A, Ajalloeian R, Hajian A (2023) ANN and MANFIS to predict pressuremeter modulus and limit pressure, case study: Isfahan metro line 2. Arab J Geosci 16(2):104
32. Sharif SI et al (2021) Application of artificial neural networks (ANNs) to predict pressure meter modulus using geotechnical and geophysical parameters. In: Symposium on the application of geophysics to engineering and environmental problems 2021. Society of Exploration Geophysicists and Environment and Engineering

33. Koza JR (1992) Genetic programming: on the programming of computers by means of natural selection. MIT Press, Cambridge, MA
34. Koza JR (1994) Genetic programming as a means for programming computers by natural selection. Stat Comput 4(2):87–112
35. Ferreira C (2001) Gene expression programming: a new adaptive algorithm for solving problems. arXiv preprint cs/0102027
36. Ferreira C (2006) Gene expression programming: mathematical modeling by an artificial intelligence, vol 21. Springer
37. Alavi AH et al (2013) Design equations for prediction of pressuremeter soil deformation moduli utilizing expression programming systems. Neural Comput Appl 23(6):1771–1786
38. Briaud J-L, Li Y, Rhee K (2006) BCD: a soil modulus device for compaction control. J Geotech Geoenviron 132(1):108–115
39. Briaud J-L (2001) Introduction to soil moduli. Geotech News 19(2):54–58
40. Reznik YM (1995) Comparison of results of oedometer and plate load tests performed on collapsible soils. Eng Geol 39(1–2):17–30
41. Reznik YM (2007) Influence of physical properties on deformation characteristics of collapsible soils. Eng Geol 92(1–2):27–37
42. Mollahasani A, Alavi AH, Gandomi AH (2011) Empirical modeling of plate load test moduli of soil via gene expression programming. Comput Geotech 38(2):281–286
43. Banzhaf W et al (1998) Genetic programming: an introduction: on the automatic evolution of computer programs and its applications. Morgan Kaufmann Publishers Inc.
44. Wang H-L, Yin Z-Y (2020) High performance prediction of soil compaction parameters using multi expression programming. Eng Geol 276:105758
45. Alavi AH, Gandomi AH (2011) A robust data mining approach for formulation of geotechnical engineering systems. Eng Comput 28:242
46. Shahin M, Jaksa M (2005) Neural network prediction of pullout capacity of marquee ground anchors. Comput Geotech 32(3):153–163
47. Mohammadzadeh D, Bazaz JB, Alavi AH (2014) An evolutionary computational approach for formulation of compression index of fine-grained soils. Eng Appl Artif Intell 33:58–68
48. Keedwell E, Khu S-T (2005) A hybrid genetic algorithm for the design of water distribution networks. Eng Appl Artif Intell 18(4):461–472
49. de Castilho VC, El Debs MK, do Carmo Nicoletti M (2007) Using a modified genetic algorithm to minimize the production costs for slabs of precast prestressed concrete joists. Eng Appl Artif Intell 20(4):519–530
50. Sanchis J et al (2010) Modelling preferences in multi-objective engineering design. Eng Appl Artif Intell 23(8):1255–1264
51. Cheng Z-L, Zhou W-H, Garg A (2020) Genetic programming model for estimating soil suction in shallow soil layers in the vicinity of a tree. Eng Geol 268:105506
52. Valatsos P et al (2021) Freight transportation route time prediction with ensemble learning techniques. In: Proceedings of the 25th Pan-Hellenic conference on informatics
53. Baykasoğlu A et al (2008) Prediction of compressive and tensile strength of limestone via genetic programming. Expert Syst Appl 35(1–2):111–123
54. Zhang XY et al (2015) Sobol sensitivity analysis: a tool to guide the development and evaluation of systems pharmacology models. CPT Pharmacometrics Syst Pharmacol 4(2):69–79

Chapter 8
Machine Learning Analysis of Lake Van and Lake Urmia Using Satellite Data

Mozhgan Yarahmadi, Mahmood Rahmani Firozjaei, and Mahdieh Kalhori

Abstract As vital components of the hydrological cycle and ecological balance, lakes play a crucial role in maintaining environmental stability. The significance of these aquatic ecosystems underscores the necessity for continuous monitoring and detailed analysis of their dynamics to achieve sustainable water resource management and balance economic-environmental considerations in watershed areas. This study utilizes remote sensing capabilities and geographic information systems to examine changes in Lake Urmia's surface area over a 33-year period (1990–2023). Landsat satellite imagery, spectral indices (NDVI and MNDWI), and climate data including annual precipitation and temperature were employed as key variables influencing the lake's condition. Based on comprehensive analyses, predictions of future water level trends were generated using the Random Forest machine learning algorithm. Findings reveal a substantial and continuous decline in the lake's water level throughout the study period. This reduction coincided with fluctuations in surrounding vegetation cover, primarily driven by agricultural expansion and excessive groundwater extraction. While climatic data analysis indicates relatively stable precipitation patterns in the lake's basin, temperature data show a gradual but significant increasing trend. These observations clearly emphasize the dominant role of human activities and inefficient water resource management in the lake's depletion. Projections for the 2024–2032 period indicate a continuing decline in water levels under current water resource utilization patterns. These findings highlight the urgent need for integrated water resource management approaches, protection of feeder watersheds, implementation of sustainable land use planning, and establishment of continuous monitoring systems to prevent irreversible degradation of this valuable ecosystem.

M. Yarahmadi (✉)
Watershed Management, Faculty of Natural Resources, University of Tehran, Tehran, Iran

M. R. Firozjaei
Faculty of Civil Engineering, College of Engineering, University of Tehran, Tehran, Iran

M. Kalhori
Department of Civil Engineering, University of Qom, Qom, Iran
e-mail: M.Kalhori@stu.qom.ac.ir

M. Mirrashid et al. (eds.), *Intelligent Predictive Systems*, Emerging Trends in Mechatronics, https://doi.org/10.1007/978-981-95-7766-8_8

Keywords Lake Urmia · Remote sensing · NDVI · MNDWI · Random Forest (RF) · Water body change

8.1 Introduction

Wetlands rank among the most valuable ecosystems on Earth, often referred to as the "kidneys of the planet" due to their crucial functions in environmental filtration and water cycle regulation [1–4]. These unique biomes play an irreplaceable role in moderating climatic conditions and conserving water resources. Scientific evidence indicates that climate change and human activities represent the two primary drivers of wetland degradation worldwide [5, 6]. Significant concern exists regarding the capacity of these ecosystems for nonlinear and abrupt changes, which could carry serious consequences for human welfare.

Numerous studies from various global regions, particularly Iran, have reported that many wetlands have diminished in area or completely disappeared due to continuous human interventions including land use changes [7–9]. Traditional methods for monitoring wetland changes primarily relied on field surveys and visual interpretation of aerial photographs [10, 11]. However, these approaches typically prove costly and time-consuming while offering limited spatial coverage, thus demonstrating reduced efficiency for large-scale or long-term monitoring [12, 13]. Recent advances in remote sensing technologies and geographic information systems have enabled more effective and comprehensive monitoring of these environments [14]. These technologies provide reliable information about land cover and establish a solid scientific and technical foundation for assessing land surface changes, particularly in wetland areas.

In recent years, increasing access to multi-satellite and multispectral remote sensing data across various spatial and temporal scales has created valuable opportunities for wetland monitoring at national and global levels [15, 16]. Among these datasets, MODIS, Sentinel, and Landsat data see the most frequent application. Although MODIS data feature relatively low spatial resolution and Sentinel lacks sufficient temporal depth for long-term analyses, the Landsat program offers an optimal balance with its 30-m resolution and rich historical archive dating back to 1990 [17]. These characteristics make Landsat imagery particularly effective for tracking long-term land surface dynamics.

Multi-temporal datasets from Landsat's Thematic Mapper (TM), Enhanced Thematic Mapper Plus (ETM+), and Operational Land Imager (OLI) sensors have been widely employed for analyzing spatiotemporal changes in lakes and water bodies [18]. Currently, medium spatiotemporal resolution imagery such as Sentinel-2 and Landsat-8 OLI are extensively used for monitoring land resources [19].

These multispectral datasets enable researchers to analyze land surface dynamics using spectral indices. Among the most widely applied indices are the Normalized

Difference Water Index (NDWI) and the Normalized Difference Vegetation Index (NDVI). For instance, MNDWI and NDVI indices were utilized to examine wetland changes in the Renuka Wetland [20].

The Random Forest algorithm, as a powerful machine learning method, has gained extensive application in environmental and remote sensing studies due to its capability to handle complex and nonlinear datasets [21–23]. Integrating vegetation indices (NDVI) with water-related indices (MNDWI) enables accurate prediction of wetland surface water dynamics. This approach serves as an effective tool for water resource management and wetland ecosystem protection by enhancing classification accuracy [24, 25].

Predictions for Lake Urmia using the Random Forest model indicate that the water body extent will undergo significant reduction by 2035 [26]. Building upon the reviewed literature, the present study aims to predict changes in the water extent of Lake Urmia from 1990 to 2023 using remote sensing data, NDVI and MNDWI spectral indices, and climatic variables including precipitation and temperature. This approach provides a clear understanding of the ecological dynamics of this crucial water body and establishes a foundation for effective planning in water resource management and conservation of its dependent ecosystems.

8.2 Materials and Methods

8.2.1 Study Area

Lake Urmia, spanning approximately 5050 km^2, holds distinction as Iran's largest inland lake and ranks as the world's second-largest hypersaline lake [27]. This valuable water body is geographically positioned between 45°–46° E longitude and 37°–38.5° N latitude. The lake's watershed extends across about 51,110 km^2 in northwestern Iran, encompassing parts of East Azerbaijan, West Azerbaijan, and Kurdistan provinces [28].

Physically, Lake Urmia measures approximately 120 km in length with an average width of 50 km. Elevation across the basin varies significantly from 1215 to 3055 m above sea level, while the lake maintains an average depth of around 18 m. Climatically, the watershed receives mean annual precipitation of approximately 355 mm, with annual temperatures fluctuating between 6.5 and 13.5 °C. The Urmia basin experiences substantial influences from both natural processes and anthropogenic activities. The hydrological network comprises approximately 39 operational dams alongside 15 dams currently under construction. Notable hydraulic structures include the Mahabad Dam, which supplies inflows to the lake; the Boukan Dam, supporting agricultural development and drinking water supply; the Zola Dam, providing water for local rivers; and the Darreh Ziyad Dam, designed for basin water regulation and management.

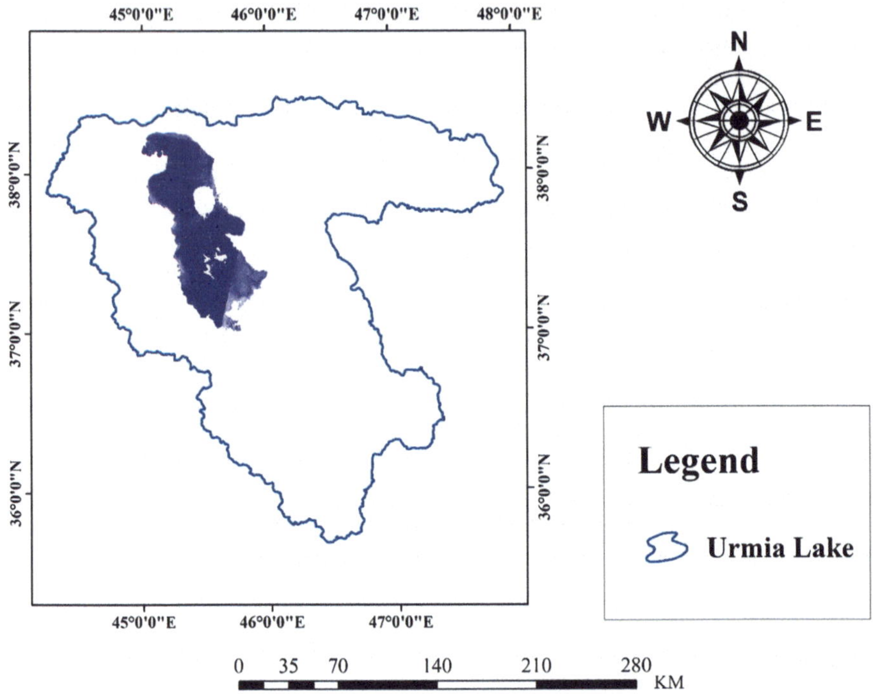

Fig. 8.1 Study area (Urmia Lake)

Prominent rivers within the basin, including the Zarineh Roud, Simineh Roud, and Ajichay, play crucial roles in maintaining the lake's water balance and ecological stability. Figure 8.1 illustrates the geographical location of the study area [27, 28].

8.2.2 Research Methodology

This research conducted a detailed examination of water extent changes in Lake Urmia across five temporal benchmarks: 1990, 2013, 2016, 2019, and 2023. The investigation employed Geographic Information Systems (GIS) and remote sensing data to precisely identify and document historical fluctuations in the lake's areal extent. Building upon the collected data, predictive analyses were also developed for the lake's water extent from 2025 to 2032, enabling identification of potential future trends.

To achieve these objectives, multispectral satellite imagery from Landsat5 and Landsat8 platforms was utilized. Table 8.1 presents the technical and temporal specifications of these images. All required imagery was acquired from the United States Geological Survey's Earth Explorer portal (earthexplorer.usgs.gov). Following

Table 8.1 Specifications of the satellite images used

Date	Satellite	Sensor	Path/row	Resolution
23/06/1990	Landsat5	TM	167/34, 167/35, 169/34	30
06/06/2013	Landsat5	TM	167/34, 167/35, 169/34	30
21/06/2016	Landsat5	TM	167/34, 167/35, 169/34	30
23/06/2019	Landsat8	OLI	167/34, 167/35, 169/34	30
17/05/2023	Landsat8	OLI	167/34, 167/35, 169/34	30

download, the images underwent comprehensive preprocessing and preparation procedures to ensure data reliability and analytical accuracy.

For generating precise water extent change maps, both radiometric and atmospheric corrections were implemented on the satellite imagery. These corrections significantly enhanced data quality and facilitated more accurate information extraction. Furthermore, the NDVI and NDWI indices were applied to assess vegetation coverage and water body extent, respectively. These indices, calculated according to established formulas referenced in [29, 30], played a pivotal role in analyzing environmental changes within the lake's boundaries and surrounding areas (Eqs. 8.1 and 8.2).

Normalized Difference Vegetation Index (NDVI)

$$\mathrm{NDVI} = \frac{\mathrm{NIR} - \mathrm{R}}{\mathrm{NIR} + \mathrm{R}} \tag{8.1}$$

Normalized Difference Water Index (NDWI)

$$\mathrm{MNDWI} = \frac{\mathrm{NIR} - \mathrm{SWR}}{\mathrm{NIR} + \mathrm{SWR}} \tag{8.2}$$

In these equations, NIR represents the near-infrared band, R represents the red band, and SWR corresponds to the shortwave infrared band. The range of values for both indices varies between −1 and 1, where values closer to 1 indicate higher index values.

8.2.3 Accuracy Assessment

Following the generation of lake water extent maps, the derived results underwent quantitative evaluation using standard accuracy assessment criteria to verify their reliability and precision. To achieve this objective, the classified water extent maps extracted from satellite imagery were cross-referenced with authentic and reliable environmental data, particularly information gathered through field surveys. This comparative analysis facilitated meticulous examination of the correspondence between the mapped data and actual ground conditions. Subsequently, an error matrix, commonly referred to as a confusion matrix, was constructed to enable more precise analysis of discrepancies and potential classification errors. This matrix permitted enhanced evaluation of classification quality and identification of errors within individual classes. Utilizing the matrix data, various metrics—including overall accuracy, kappa coefficient, user accuracy, and producer accuracy—were computed to deliver a comprehensive assessment of the water extent maps' accuracy. This evaluation methodology adheres to the approach established by Kafy et al. (2020), which is widely recognized as a standard protocol in remote sensing data analysis.

8.2.4 Random Forest

The Random Forest algorithm represents a widely adopted machine learning method for classification and prediction tasks. This approach operates on the foundation of decision trees, wherein multiple independent decision trees are generated, each trained on a random subset of the data. Following the training phase, outcomes from all trees are aggregated to produce final predictions. Among the algorithm's prominent advantages are its capacity to handle incomplete or imbalanced datasets and mitigate issues such as arbitrary variable selection that may occur in other models. Furthermore, by simultaneously training numerous trees on random data portions, the risk of overfitting is diminished, and the model acquires enhanced generalization capability for new data. Access to substantial volumes of high-quality data plays a decisive role in training machine learning and artificial intelligence models. Employing appropriate data enhances model accuracy and effectiveness in addressing complex problems. Integrating diverse data collection and processing methodologies not only improves model precision but also fosters creativity and innovation in interdisciplinary research [31–33]. Due to these distinctive characteristics, Random Forest is recognized as one of the most efficient and robust machine learning algorithms, finding extensive applications in automatic classification, data forecasting, and supervised learning tasks. For these reasons, the current research employs this method to predict water area changes in Lake Urmia.

8.3 Results and Discussion

8.3.1 Water Extent and Accuracy Assessment

Analysis of the water extent mapping results for Lake Urmia in 1990, 2013, 2016, 2019, and 2023 demonstrates high accuracy across the study area. The 1990 water extent classification achieved 99% overall accuracy with a kappa coefficient exceeding 96%. In 2013, these metrics were recorded at 99% and 97%, respectively. Both the 2016 and 2019 maps demonstrated 98% overall accuracy and 98% kappa coefficient. Finally, the 2023 classification attained 99% for both overall accuracy and kappa coefficient.

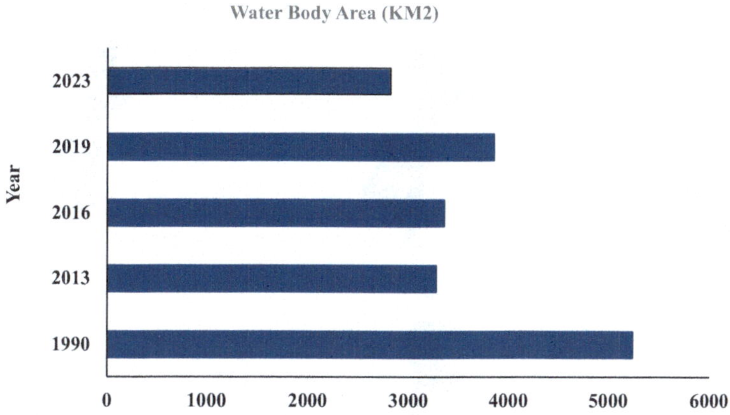

Fig. 8.2 Changes in the water surface area of Lake Urmia

The estimated changes in Lake Urmia's water surface area between 1990 and 2023 reveal significant fluctuations. The lake covered approximately 5443.38 km^2 in 1990, which decreased to about 3277.44 km^2 by 2013. This was followed by an increase to 3357.55 km^2 in 2016 and further expansion to 3855.79 km^2 in 2019. However, by 2023, the water area had experienced a notable decline to 2820.17 km^2 (Fig. 8.2).

8.3.2 MNDWI Index

The variations in the Modified Normalized Difference Water Index (MNDWI) for Lake Urmia are presented in Fig. 8.3. As illustrated in the graph displaying the average MNDWI values (Fig. 8.4), the highest average index value of 0.36 was recorded in 2023, indicating significant alterations in the spectral characteristics of the lake's water surface during this period.

8.3.3 Temporal Changes in MNDWI

Analysis of Lake Urmia's water surface area changes across different periods reveals considerable fluctuations. The obtained data indicate that during the 1990–2013 period; the lake experienced a 19.92% reduction in its water area. In the 2013–2016 interval, while the lake gained 2609.64 km^2 in surface area, it simultaneously lost 1862.415 km^2. These changes manifested as a 54.84% increase and 45.16% decrease during the 2016–2019 period.

In the most recent period (2019–2023), the lake's water area demonstrates a reduction of 1213.17 km^2 alongside an increase of 1597.16 km^2. Overall, across the

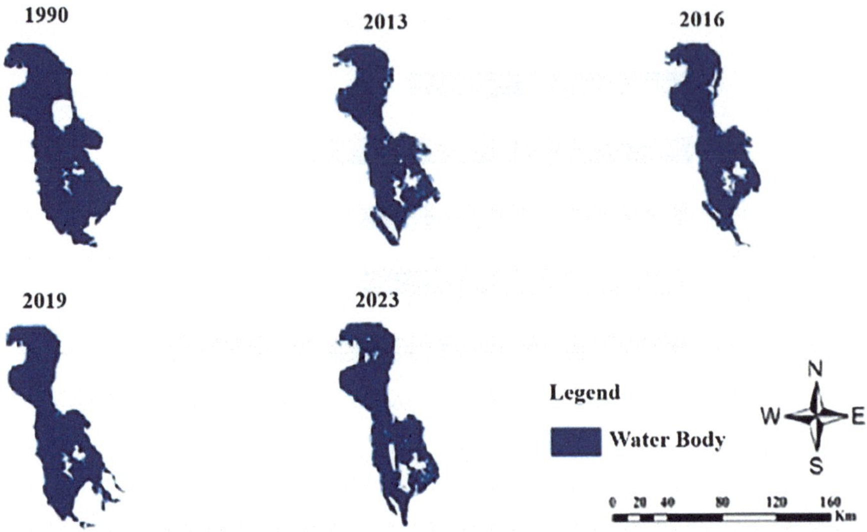

Fig. 8.3 MNDWI index of the Lake Urmia region, Iran

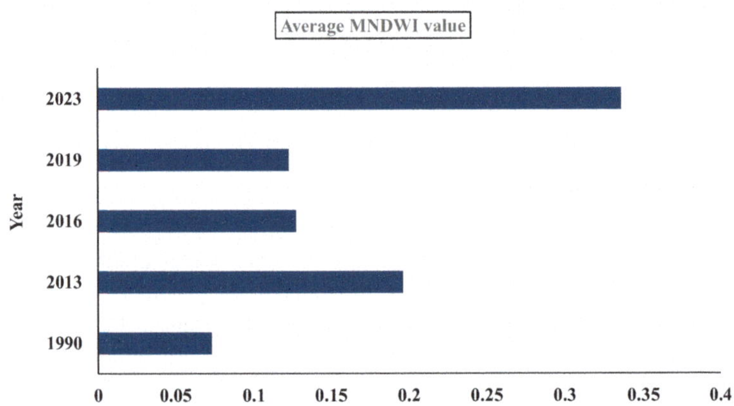

Fig. 8.4 Temporal variations in the average MNDWI index

33-year timeframe from 1990 to 2023, Lake Urmia has undergone a 43.05% decrease and 56.95% increase in its water surface area (Fig. 8.5 and Table 8.2).

According to previous research, Lake Urmia shrank to approximately one-third of its original size between 1984 and 2018 [28]. Similarly, during the 1970–2020 period, about 20–50% of the lake's surface water has been lost. Due to the combined effects of climate change and water resource development, the annual inflow into Lake Urmia has decreased by 48% compared to previous years [34].

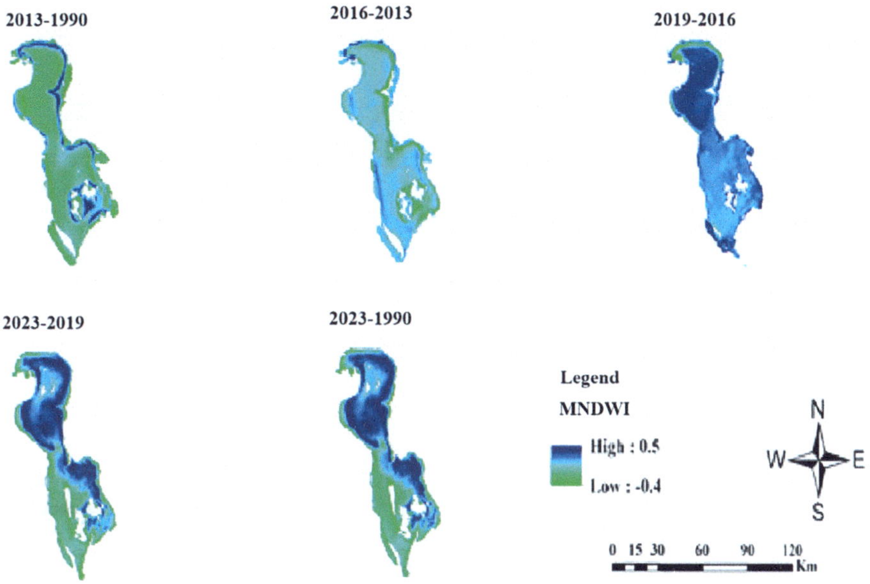

Fig. 8.5 Yearly variations in the MNDWI index in the Lake Urmia region, Iran

Table 8.2 Trends of Increase and decrease in Lake Urmia

Urmia			
Year	Trend	Decreasing	Increasing
2013–1990	%	92.19	80.08
	KM2	16.651	46.2618
2016–2013	%	73.13	27.86
	KM2	19.415	64.2609
2019–2016	%	16.45	84.54
	KM2	86.1482	77.18
2023–2019	%	16.43	84.56
	KM2	17.1213	61.1597
2023–1990	%	43.05	95.56
	KM2	24.1211	32.1602

Studies further indicate that within the Lake Urmia basin, over 56% of agricultural water demand is met from surface water sources, while approximately 44% is supplied from groundwater resources [35]. With the continuous decline in the lake's water level and diminishing upstream inflows, dependence on groundwater for agricultural purposes has intensified, frequently through the drilling of unauthorized wells.

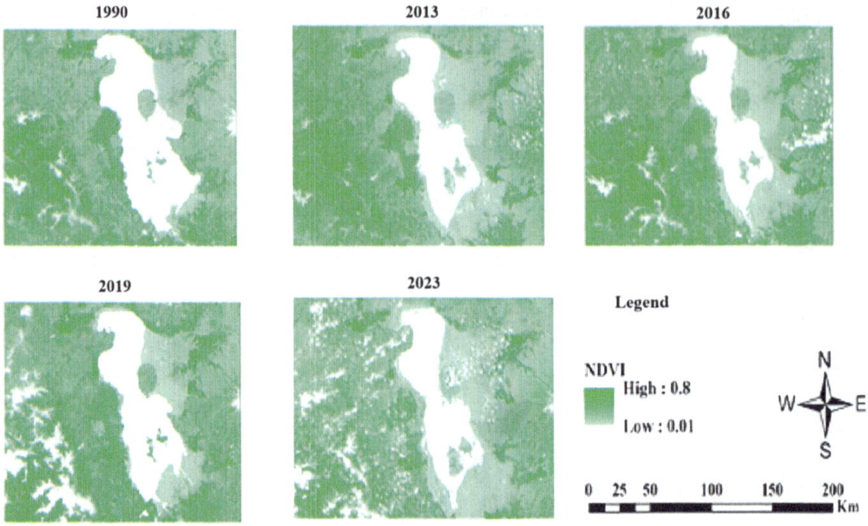

Fig. 8.6 Changes in the NDVI index in the Lake Urmia region, Iran

8.3.4 NDVI Index

Analysis of the Normalized Difference Vegetation Index (NDVI) within the Lake Urmia basin (Fig. 8.6) has revealed considerable fluctuations throughout the research period. Spatially, the highest NDVI values were detected in the western sectors of the lake, whereas the lowest values were recorded in the eastern and southeastern regions. Examination of NDVI data between 2000 and 2018 demonstrates that despite substantial reduction in water resources, vegetation coverage experienced periodic increases [36]. The research further established that desertification processes, driven by both natural determinants and anthropogenic activities, have significantly contributed to ecosystem degradation and vegetation loss in the region.

8.3.5 Trends in Vegetation Cover (NDVI) in the Lake Urmia Basin

Analysis of the Normalized Difference Vegetation Index (NDVI) in the Lake Urmia Basin (Fig. 8.7) over the 1990–2023 period reveals that despite substantial reduction in the lake's water surface area, the index demonstrated increasing trends during specific intervals. Temporal data indicate that NDVI values rose from approximately 0.16 in 1990 to 0.26 in 2013, followed by fluctuating declines in subsequent years, eventually reaching around 0.15 in 2023.

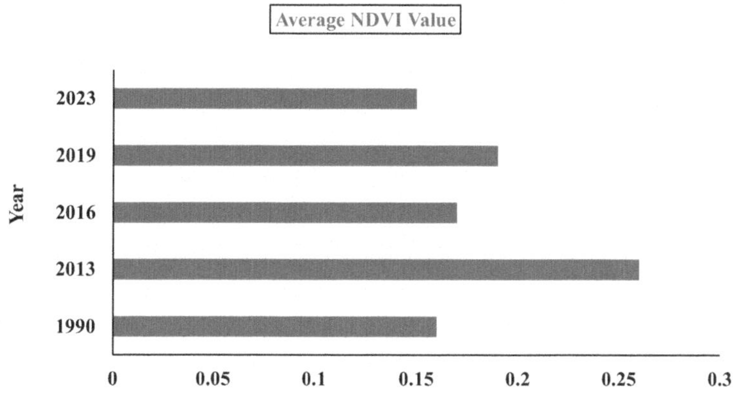

Fig. 8.7 Temporal variations in the average NDVI index

This paradoxical pattern suggests that while the lake's area has diminished, expanded agricultural activities within the basin have stimulated greater vegetation growth, consequently elevating NDVI values during certain periods. In other words, the observed NDVI increase does not necessarily signal ecological improvement but rather reflects the expansion of water-intensive crop cultivation and extensive exploitation of surface and groundwater resources [37]. The stark contrast between the persistent reduction in lake area and relative vegetation increase in some years underscores unsustainable water resource management in the Urmia Basin. Areas previously constituting part of the lake's water body have been converted to agricultural lands or pastures. This trajectory not only hinders the lake's recovery but also intensifies pressure on water resources and exacerbates the drought crisis [38].

8.4 Climatic Changes in Lake Urmia

8.4.1 Annual Precipitation Trends

Analysis of annual precipitation patterns in the Lake Urmia basin demonstrates that throughout the nearly three-decade study period, rainfall has exhibited minor fluctuations while maintaining relative stability. The highest annual precipitation values ranged between 282 and 285 mm, with the lowest average recorded at 281.88 mm and the highest at 285.10 mm. This narrow variability range indicates that contrary to common perceptions, the sharp decline in Lake Urmia's water level cannot be attributed to reduced precipitation. Across different periods, precipitation changes have generally remained consistent. For instance, certain years such as 2010 witnessed slight precipitation increases (approximately 284.9 mm), but subsequent years returned to long-term averages. This fluctuating yet stable pattern reveals that regional climate conditions regarding precipitation have not undergone

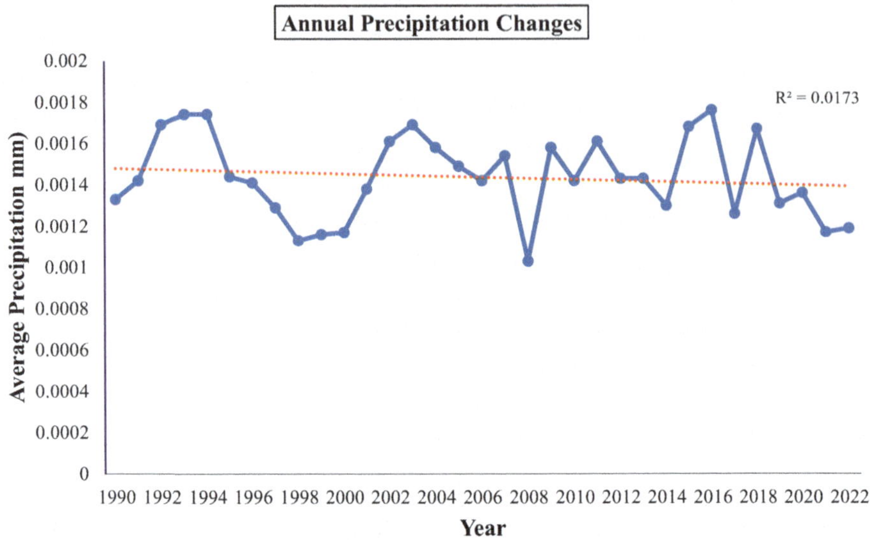

Fig. 8.8 Annual precipitation variations in the Lake Urmia basin

fundamental transformations. Comparing precipitation trends with changes in the lake's water extent reveals a crucial insight: while the lake's area has dramatically diminished from the 1990s to the 2020s, with projections indicating near-complete desiccation by 2025, rainfall has remained virtually unchanged. This absence of direct correlation between precipitation and water area underscores the dominant influence of human activities, including excessive groundwater extraction, water-intensive agricultural expansion, and construction of numerous dams on inflow rivers [39, 40].

Consequently, annual rainfall data emphasize that Lake Urmia's desiccation primarily results from unsustainable water resource management and anthropogenic pressures rather than climatic variations or precipitation reduction. These findings highlight the urgent necessity for comprehensive planning and intervention in regional water resource management (Fig. 8.8).

8.4.2 Changes in Average Temperature

Analysis of annual temperature data in the Lake Urmia basin, as illustrated in Fig. 8.9, reveals an overall warming trend from 1990 to 2023. Temperature values demonstrated minor fluctuations during this period, with the mean temperature decreasing from 0.00133 in 1990 to 0.00119 in 2023.

During the 1990–2001 interval, the basin exhibited a positive temperature trend with an estimated annual slope of 0.00035 and a coefficient of determination (R^2) of 0.48, indicating a relatively strong warming pattern throughout this decade. The temperature rise during this period likely correlates with diminished surface water

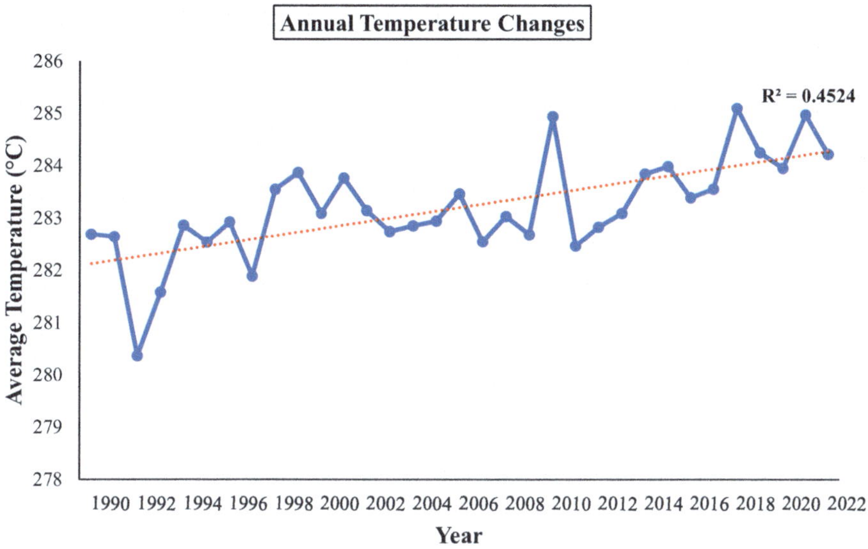

Fig. 8.9 Annual temperature variations in the Lake Urmia basin

resources and enhanced evaporation, consequently accelerating the lake's desiccation. From 2001 to 2012, temperature variations remained relatively stable, with average values fluctuating between 0.00116 and 0.00161, suggesting continued warming compared to the previous decade. However, minimum and maximum temperatures during this interval reached 0.00113 and 0.00169, respectively, reflecting seasonal and interannual variability.

Between 2012 and 2022, the temperature trend underwent substantial renewed increase. The change slope was positively significant, with values ranging from 0.00119 to 0.00176, indicating elevated evaporation rates and the impact of higher temperatures on the reduction of the lake's water area. These findings demonstrate that despite seasonal and decadal fluctuations, the annual temperature increase in the Urmia Basin—coupled with reduced surface water flow and enhanced evaporation—has constituted a key factor in the lake's drying [41, 42]. Consequently, water resource management strategies should incorporate temperature trends and their effects on basin drought conditions to mitigate the environmental and economic consequences associated with lake water depletion.

8.4.3 Prediction of Water Extent Changes in Lake Urmia Using RF

Trend analysis of data spanning 1990 to 2023 has revealed a persistent decline in Lake Urmia's surface area, diminishing from approximately 5000 km^2 to under 3000 km^2. Model projections for the 2025–2032 period indicate continued

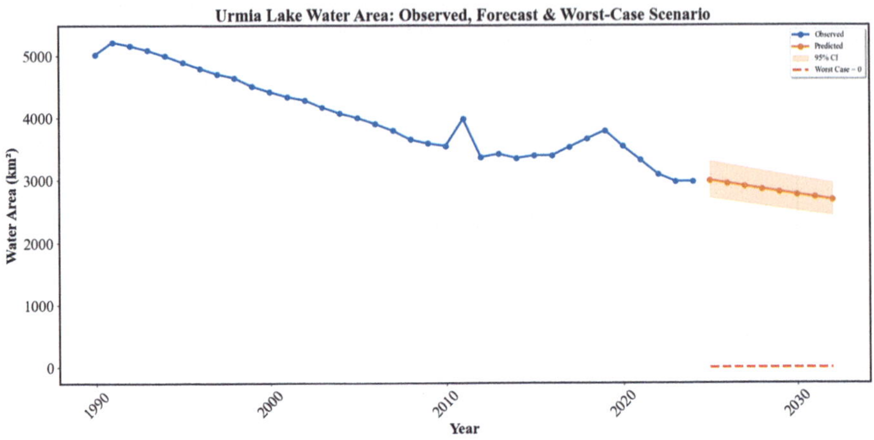

Fig. 8.10 Predicted water extent changes of Lake Urmia from 2024 to 2032

reduction in the lake's areal extent though at a more moderate rate compared to previous decades. The 95% confidence interval delineates the spectrum of potential changes, reflecting inherent uncertainties embedded within such prognostications. For sensitivity and criticality assessments, a pessimistic scenario was evaluated presuming the lake's surface area would diminish to zero from 2025 onward. Although this scenario represents an extreme circumstance, it corresponds with historical declining trends and accentuates the lake's pronounced vulnerability to climate change and water resource exploitation [43, 44].

The outcomes demonstrate that under critical conditions, complete desiccation of the lake remains plausible in the short term. Nevertheless, the forecasting model still predicts a gradual and moderated recession in the lake's water area. These findings underscore the pressing necessity for implementing effective water resource management and lake restoration strategies, thereby establishing a foundation for informed decision-making and long-term policy formulation (Fig. 8.10).

8.5 Conclusion

The results derived from MNDWI and NDVI indices, combined with Random Forest model projections, demonstrate a substantial and persistent decline in Lake Urmia's water volume in recent decades. The high precision in mapping water level changes and validation of these indices confirm data reliability and provide an accurate representation of the region's ecological status. Although vegetation coverage increases have been observed, these changes primarily stem from agricultural expansion and intensive groundwater extraction rather than natural improvement in the lake's condition. Regional climatic trends, including minor precipitation variations and gradual temperature increases, indicate that the severe lake surface reduction results mainly from anthropogenic pressures and inefficient water resource management rather than natural factors. Excessive irrigation, water-intensive agriculture, and construction of numerous dams have served as primary drivers of the lake's desiccation.

These findings reveal a reinforcing cycle wherein depletion of surface and groundwater resources, intensified agricultural exploitation, and human-induced environmental pressures mutually amplify each other, accelerating the lake's recession. Model projections indicate this declining trend will persist, with complete drying feasible in the short term under pessimistic scenarios. The outcomes emphasize the urgent requirement for effective management interventions including curbing excessive water abstraction (groundwater and unauthorized wells), protecting recharge zones, optimizing land use, and implementing continuous monitoring. Such measures prove essential for mitigating environmental and socio-economic impacts, maintaining residual ecosystem functions, and ensuring sustainable water resource management within the basin.

Collectively, these findings establish a scientific foundation for informed decision-making and long-term policy development in water resource management and ecosystem protection, underscoring that a combination of preventive management strategies and continuous monitoring remains crucial to prevent further environmental degradation and safeguard the socio-ecological integrity of Lake Urmia.

AI Use Disclosure Statement The authors used generative AI technologies only for proofreading. This includes minor corrections to grammar, spelling, and syntax. No AI was used for ideation, data analysis, or writing original content.

References

1. Bertassello LE, Basu NB, Maes J, Grizzetti B, La Notte A, Feyen L (2025) The important role of wetland conservation and restoration in nitrogen removal across European river basins. Nat Water 3(8):867–880
2. Song F, Su F, Mi C, Sun D (2021) Analysis of driving forces on wetland ecosystem services value change: a case in Northeast China. Sci Total Environ 751:141778. https://doi.org/10.1016/j.scitotenv.2020.141778
3. Zhang X, Wang G, Xue B, Zhang M, Tan Z (2021) Dynamic landscapes and the driving forces in the Yellow River Delta wetland region in the past four decades. Sci Total Environ 787:147644. https://doi.org/10.1016/j.scitotenv.2021.1
4. Wang Y, Zhang P, Sun H, Jia X, Zhang C, Liu S, Shao MA (2022) Vertical patterns and controlling factors of soil nitrogen in deep profiles on the Loess Plateau of China. Catena 215:106318
5. Shi H, Lu J, Zheng W, Sun J, Li J, Guo Z et al (2020) Evaluation system of coastal wetland ecological vulnerability under the synergetic influence of land and sea: a case study in the Yellow River Delta, China. Mar Pollut Bull 161:111735
6. Singh S, Bhardwaj A, Verma VK (2020) Remote sensing and GIS based analysis of temporal land use/land cover and water quality changes in Harike wetland ecosystem, Punjab, India. J Environ Manag 262:110355. https://doi.org/10.1016/j.jenvman.2020.110355
7. Purnama F, Ramli I, Dahlan D (2025) Analysis of land use/land cover change (LULC) Paya Nie wetland using remote sensing method. In: IOP conference series: earth and environmental science, vol 1510(1). IOP Publishing, p 012014

8. Mao D, Wang Z, Du B, Li L, Tian Y, Jia M, Zeng Y, Song K, Jiang M, Wang Y (2020) National wetland mapping in China: a new product resulting from object-based and hierarchical classification of Landsat 8 OLI images. ISPRS J Photogramm Remote Sens 164:11–25
9. Lu M, Zou Y, Xun Q, Yu Z, Jiang M, Sheng L, Lu X, Wang D (2021) Anthropogenic disturbances caused declines in the wetland area and carbon pool in China during the last four decades. Glob Change Biol 27(16):3837–3845
10. Mu S, Li B, Yao J, Yang G, Wan R, Xu X (2020) Monitoring the spatio-temporal dynamics of the wetland vegetation in Poyang Lake by Landsat and MODIS observations. Sci Total Environ 725:138096
11. Mao D, Wang Z, Wu J, Wu B, Zeng Y, Song K, Yi K, Luo L (2018) China's wetlands loss to urban expansion. Land Degrad Dev 29(8):2644–2657
12. Qureshi S, Alavipanah SK, Konyushkova M, Mijani N, Fathololomi S, Firozjaei MK, Homaee M, Hamzeh S, Kakroodi AA (2020) A remotely sensed assessment of surface ecological change over the Gomishan wetland, Iran. Remote Sens 12(18):2989
13. Stephenson P, Ntiamoa-Baidu Y, Simaika JP (2020) The use of traditional and modern tools for monitoring wetlands biodiversity in Africa: challenges and opportunities. Front Environ Sci 8:61
14. Gong M, Wu L, Bi XY, Ren LM, Wang L, Ma ZD et al (2010) Assessing heavy-metal contamination and sources by GIS-based approach and multivariate analysis of urban–rural topsoils in Wuhan, central China. Environ Geochem Health 32(1):59–72
15. Yuan S, Liang X, Lin T, Chen S, Liu R, Wang J et al (2025) A comprehensive review of remote sensing in wetland classification and mapping. arXiv preprint arXiv:2504.10842
16. Hou M, Ge J, Gao J, Meng B, Li Y, Yin J, Liu J, Feng Q, Liang T (2020) Ecological risk assessment and impact factor analysis of alpine wetland ecosystem based on LUCC and boosted regression tree on the Zoige Plateau, China. Remote Sens 12(3):368
17. Pekel J-F, Cottam A, Gorelick N, Belward AS (2016) High-resolution mapping of global surface water and its long-term changes. Nature 540(7633):418–422
18. Kirtiloglu O, Orhan O, Ekercin S (2016) A map mash-up application: investigation the temporal effects of climate change on salt lake basin. Int Arch Photogramm Remote Sens Spatial Inf Sci 41. https://doi.org/10.5194/isprs-archives-XLI-B4-221-2016
19. Roy PS, Behera MD, Srivastav SK (2017) Satellite remote sensing: sensors, applications and techniques. Proc Natl Acad Sci India A Phys Sci 87(4):465–472
20. Ashok A, Rani HP, Jayakumar KV (2021) Monitoring of dynamic wetland changes using NDVI and NDWI based landsat imagery. Remote Sens Appl Soc Environ 23:100547. https://doi.org/10.1016/j.rsase.2021.100547
21. Savitha C, Talari R (2023) Mapping cropland extent using Sentinel-2 datasets and machine learning algorithms for an agriculture watershed. Smart Agric Technol 4:100193
22. Sun Y, Li Y, Wang R, Ma R (2024) Modelling potential land suitability of large-scale wind energy development using explainable machine learning techniques: applications for China, USA and EU. Energy Convers Manag 302:118131
23. Tiwari P, Poudel KP, Yang J, Silva B, Yang Y, McConnell M (2023) Marginal agricultural land identification in the lower Mississippi Alluvial Valley based on remote sensing and machine learning model. Int J Appl Earth Obs Geoinf 125:103568
24. Indriani RF, Anjasmara IM, Utama W, Paramita EGK, Nainggolan RAO (2023) Comparative analysis of physiograpic study for hydrology of Benowo Region, Surabaya. In: IOP conference series: earth and environmental science, vol 1250(1). IOP Publishing, p 012015
25. Chowdhury MS (2024) Comparison of accuracy and reliability of random forest, support vector machine, artificial neural network and maximum likelihood method in land use/cover classification of urban setting. Environ Challenges 14:100800
26. Sakizadeh M, Milewski A (2024) Quantifying LULC changes in Urmia Lake Basin using machine learning techniques, intensity analysis and a combined method of cellular automata (CA) and artificial neural networks (ANN) (CA-ANN). Model Earth Syst Environ 10(2):2011–2030
27. Eimanifar A, Mohebbi F (2007) Urmia Lake (northwest Iran): a brief review. Saline Syst 3(1):5

28. Rahimi A, Breuste J (2021) Why is Urmia Lake drying up? Prognostic modeling with land-use data and artificial neural network. Front Environ Sci 9. https://doi.org/10.3389/fenvs.2021.603916
29. Tucker CJ, Justice CO, Prince SD (1986) Monitoring the grasslands of the Sahel 1984-1985. Int J Remote Sens 7(11):1571–1581
30. Gao BC (1996) NDWI—a normalized difference water index for remote sensing of vegetation liquid water from space. Remote Sens Environ 58(3):257–266
31. Gull S, Shah SR, Dar AM (2022) Assessing land use/land cover change detection of north-eastern watersheds of Kashmir valley using GIS and remote sensing techniques. Water Pract Technol 17(8):1603–1614
32. Yousefi S, Mirzaee S, Almohamad H, Al Dughairi AA, Gomez C, Siamian N, Alrasheedi M, Abdo HG (2022) Image classification and land cover mapping using Sentinel-2 imagery: optimization of SVM parameters. Land 11(7):993. https://doi.org/10.3390/land11070993
33. Svoboda J, Štych P, Laštovička J, Paluba D, Kobliuk N (2022) Random forest classification of land use, land-use change and forestry (LULUCF) using Sentinel-2 data—a case study of Czechia. Remote Sens 14(5):1189. https://doi.org/10.3390/rs14051189
34. Shadkam S, Ludwig F, van Oel P, Kirmit Ç, Kabat P (2016) Impacts of climate change and water resources development on the declining inflow into Iran's Urmia Lake. J Great Lakes Res 42(5):942–952
35. Ahmadaali J, Barani G-A, Qaderi K, Hessari B (2018) Analysis of the effects of water management strategies and climate change on the environmental and agricultural sustainability of Urmia Lake Basin, Iran. Water 10(2):160
36. Khodaei Geshlag F, Roostaei S, Mokhtari D (2020) Monitoring the desertification trend in the areas surrounding Lake Urmia (2000-2018). Geogr Environ Plan 31(3):21–40
37. Khorshiddoust AM, Patel N, Khalilzadeh E, Bostanabad SA, Tajbar S (2025) A comparative study of the surface level changes of Urmia Lake and Aral Lake during the period of 1988 to 2018 using satellite images. Front Earth Sci 19(1):108–119
38. Azizi S (2024) Drought and environmental transformations of Lake Urmia: an integrated analysis using machine learning and GIS. Eur J Appl Sci 12. https://doi.org/10.14738/aivp.125.17718
39. Kazemi Garajeh M, Akbari R, Aghaei Chaleshtori S, Shenavaei Abbasi M, Tramutoli V, Lim S, Sadeqi A (2024) A comprehensive assessment of climate change and anthropogenic effects on surface water resources in the Lake Urmia Basin, Iran. Remote Sens 16(11):1960
40. Mohammad Pourian Bazzaz P, Sadeghfam S, Khatibi R, Nourani V (2024) A drought study in the basin of Lake Urmia under climate change scenarios with higher spatial resolution to understand the resilience of the basin. J Water Clim Change 15(2):453–475
41. Ghafarian P, Tajbakhsh S, Delju AH (2021) Analysis of the long-term trend of temperature, precipitation, and dominant atmospheric phenomena in Lake Urmia. In: Lake Urmia: a hypersaline waterbody in a drying climate. Springer International Publishing, Cham, pp 149–167, 172–183
42. Radmanesh F, Esmaeili-Gisavandani H, Lotfirad M (2022) Climate change impacts on the shrinkage of Lake Urmia. J Water Clim Change 13(6):2255–2277
43. Akbari M (2023) Application of remote-sensing and machine-learning in studying the climatic and anthropogenic drivers of water bodies drying in data-scarce transboundary basins
44. Sadeghi V (2022) Monitoring and prediction of the Urmia Lake drying trend based on time-series remotely sensed images and artificial neural networks. Traitement du Signal 39(4):1229

Chapter 9
Hybridizing Deep Neural Network Models to Predict Undrained Shear Strength from CPT Data

Ali Hajiazizi, Xuzhen He, and Danial Jahed Armaghani

Abstract Many geotechnical models have been developed to predict soil behavior; however, they often fail to establish accurate correlations between design parameters and geotechnical investigations, leading to reduced prediction accuracy. Such limitations can result in technical challenges, increased construction costs, and project delays. In cohesive soils, the undrained shear strength S_u is a critical parameter influencing several geotechnical applications. This study employs a machine learning (ML) model based on the Bayesian method to predict S_u from cone penetration test (CPT) data. Models with nine and four input variables and point-estimated predictions were developed using Bayesian Deep Neural Networks (BDNN). Their performance was assessed through multiple metrics, including mean absolute error (MAE), mean squared error (MSE), R-squared score, variance accounted for (VAF), and the a20-index, achieving an R^2 of 0.814 for the training dataset and 0.869 for the testing dataset with nine inputs, and an R^2 of 0.743 for the training dataset and 0.766 for the testing dataset with four inputs. These results highlight the potential of ML, particularly Bayesian-based approaches, to improve geotechnical predictions while ensuring cost efficiency.

Keywords Deep learning · Cone penetration test · Undrained shear strength

A. Hajiazizi (✉)
School of Minerals and Energy Resources Engineering, University of New South Wales, Sydney, NSW, Australia
e-mail: a.hajiazizi@unsw.edu.au

X. He · D. Jahed Armaghani
School of Civil and Environmental Engineering, University of Technology Sydney, Sydney, NSW, Australia
e-mail: Xuzhen.He@uts.edu.au; Danial.JahedArmaghani@uts.edu.au

M. Mirrashid et al. (eds.), *Intelligent Predictive Systems*, Emerging Trends in Mechatronics, https://doi.org/10.1007/978-981-95-7766-8_9

9.1 Introduction

Undrained shear strength (S_u) indicates the ability or, in other words, the capacity of soil against shearing stress. This parameter is often considered in geotechnical design, such as soil settlement calculation and assessing slope stabilization. The heterogeneity and spatial variability of soils result in significant variations in soil parameters from site to site [1]. Phoon and Kulhawy [2] demonstrated that a single coefficient of variation (COV) is insufficient to describe soil design parameters, as they are inherently uncertain, multivariable, and often incomplete. Among these parameters, the S_u is a key indicator of soil stability. Direct (e.g., direct shear tests) and indirect methods (e.g., piezocone penetration tests) are generally used to derive internal friction and cohesion to calculate S_u. However, conventional tests encounter limitations related to tool geometry, sample quality, high costs, and delays, all of which reduce accuracy [3, 4]. To overcome these challenges, many researchers have correlated S_u with soil properties—such as water content, plasticity index, and liquidity index—as well as in situ tests including SPT, CPT, pressuremeter, vane shear, and dilatometer tests [5–9]. For instance, Rémai [5] introduced a semi-empirical method for approximating S_u using unconfined compression tests (UCT) and triaxial iso-tropically consolidated-undrained compression on Holocene clays, incorporating an empirically calculated cone factor (N_{kt}). Cheshomi et al. [6] developed three empirical equations to estimate S_u, based on 44 uniaxial and pressuremeter tests conducted on soils from Qom City, achieving an accuracy of up to 73%. Belkhatir et al. [7] investigated the effect of fine-grained content (ranging from 0% to 50%) and relative density (20% to 91%) on S_u and hydraulic conductivity in sand-silt soils from the Chelif River using a triaxial apparatus. Their results indicated that S_u decreased at the peak state with lower void ratios and higher fine-grained content up to 30%. Further increases in fine content led to a reduction in S_u at the peak state as the void ratio increased. However, the lack of comprehensive information to identify the most influential soil properties, coupled with disturbances during soil sampling, limits the applicability of these solutions. This inaccuracy impacts geotechnical analyses, such as estimating the bearing capacity of fine-grained soils, selecting appropriate soil improvement methods, and preventing structural failures in projects like dams, embankments, and tunnels. Phoon and Kulhawy [2] illustrated how site conditions, the quality of correlation models, and measurement errors in field tests contribute to the overall variability in geotechnical properties. Inherent soil variability, which is related to site conditions, is considered aleatoric in nature and cannot be reduced by additional testing. In contrast, both data points scatter around the mean trend and measurement errors are recognized as epistemic uncertainties, which can be mitigated by refining models, gathering more data, or improving testing procedures. While this error classification is helpful, it depends on the modeler's approach [8]. Generally, in situ tests are more reliable due to their capability to reflect actual field stress conditions [9]. There are three types of CPT: mechanical, electronic, and piezocone penetration tests (CPTUs). CPTUs are capable of simultaneously measuring penetration pore pressure while recording cone resistance (q_c) and skin friction (f_s). Various correlations exist between test indices, such as S_u, over-consolidation ratio (OCR), corrected cone tip resistance ($q_t = q_c + (1 - a)\ u_2$, where "a" is the area ratio of the cone), and pore water pressure behind the cone (u_2). Examples of these correlations include $S_u - q_t$ and $u_2 - q_t$ [10]. These parameters can be correlated with laboratory or field tests using theoretical, empirical, and semi-empirical models [11, 12].

Theoretical approaches, such as bearing capacity theory, form the basis of many CPT correlations, but local models may remain site-specific and limited in generality. To overcome these issues, Ching and Phoon [13] introduced a method that incorporates primary and secondary inputs, making previous models more generalized. This narrows the uncertainty, reduces biases, and globalizes the models. However, this approach does not provide solutions for its assumptions and approximations, and it inherits the errors present in other models. Additionally, the technique introduces only two new input variables—plasticity index (PI) and sensitivity (S_t)—but does not account for other important soil properties, such as preconsolidation pressure (σ_p).

Recently, soft computing approaches—such as random forest, extreme gradient boosting machine, ensemble learning, artificial neural networks (ANNs), and SVMR—have gained popularity for predicting S_u [4, 14–17] and generally in civil engineering [18–25]. Machine learning (ML) offers a promising alternative by handling multivariate and nonlinear relationships and learning directly from data. It has been applied to correlate S_u with soil parameters [14]. However, past studies often relied on a small number of datasets, used a large number of input variables without optimization, or depended heavily on experimental tests, which may have introduced systematic errors [3]. For example, Zhang et al. [16] trained three models: an ANN integrating Monte Carlo dropout (ANN_MCD), an ANN integrating variational inference (ANN_VI), and a conventional ANN. However, their dataset was relatively small (331 samples), with a testing accuracy of only 79%, and the selected model (ANN_MCD) required a higher computational cost compared to this study. Kang et al. [26] found that linear models outperformed nonlinear ones when correlating SPT with S_u in Alberta, Canada. However, their dataset was site-specific, and the best model achieved an R^2 of 0.44.

From this review, two main gaps emerge: (1) existing correlations and probabilistic models are restricted by strong assumptions, limited input parameters, and a lack of consideration of CPTU pore pressure effects; (2) prior ML models often suffer from small or site-specific datasets, a lack of uncertainty quantification, and limited feature selection, leading to poor generalization.

Given the significant variability in soil properties, Bayesian deep neural networks (BDNN), which focus on variational inference (VI), are explored to capture uncertainty and enhance prediction accuracy. The present study employs larger and more generalizable datasets, achieving an R^2 of 0.869 with nine input variables and an R^2 of 0.766 with four input variables on the testing datasets. The objectives of this research are: (1) to correlate S_u and CPT measurements applying DL models, and (2) to use multivariate features and Bayesian optimization to determine optimal hyperparameters.

9.2 Literature Review

9.2.1 Cone Penetration Test

Cone Penetration Tests (CPTs) are widely used to determine soil properties, including S_u, Liquidity Index (LI), vertical stress (σ_{v0}), pre-consolidation stress (σ_p), water content, and soil layering. The concept of CPT was first introduced in 1917. In 1932, Dutch engineer B. Barentsen developed the cone penetrometer in a form recognizable today. His initial design consisted of a 19 mm outer gas cylinder and a 15 mm inner steel rod, both fitted with a cone tip, manually inserted into the ground to depths of 10–12 m. However, the weight of the rod affected cone resistance measurements, which Barentsen later corrected by eliminating the weight effect. In 1935, T.K. Huizinga developed another version with a 10-tonne penetration rig and a 19 mm casing to reduce f_s along the inner rod. Despite this improvement, soil particles could enter the device through the gap between the casing and rod. Vermeiden and Plantema proposed a solution by adding a conical part above the cone. Over subsequent decades, CPT designs were further refined: Begemann (1953) introduced an adhesion jacket to measure skin friction and proposed the friction ratio (f_r). The first hydraulic cone penetrometer appeared in France in 1966, including manometers for steady cone resistance readings transmitted to the surface. Nevertheless, mechanical CPTs remained popular due to their simplicity, durability, and cost-effectiveness though they had limitations: data quality depended on operator skill, measurements in soft soils were unreliable, and hard strata could not be properly tested. Electrical cone penetrometers emerged during World War II in Berlin, using signal cables to transmit measurements to the surface. Key improvements included reducing inner-outer friction, enabling continuous penetration rates, improving data reliability, and allowing interpretation of soft soils using sensitive load cells. Delft Soil Mechanics Laboratory (1957) developed a version capable of measuring local side friction separately. Electrical CPTs were further advanced by Fugro and the Dutch State Research Institute in 1965, and by 1971, De Ruiter introduced electrical inclinometers to detect vertical deviations. The Norwegian Geotechnical Institute (NGI) patented an electrical piezometer for use during penetration tests. It was demonstrated that pore water pressure significantly affects data interpretation. Subsequently, instruments developed in the USA and Sweden were able to identify thin, permeable, clayey layers. Modern penetrometers are equipped with piezocones, which measure pore pressure at three specific locations: on the cone tip (u_1), just behind the cone (u_2), and behind the friction sleeve (u_3) (Fig. 9.1).

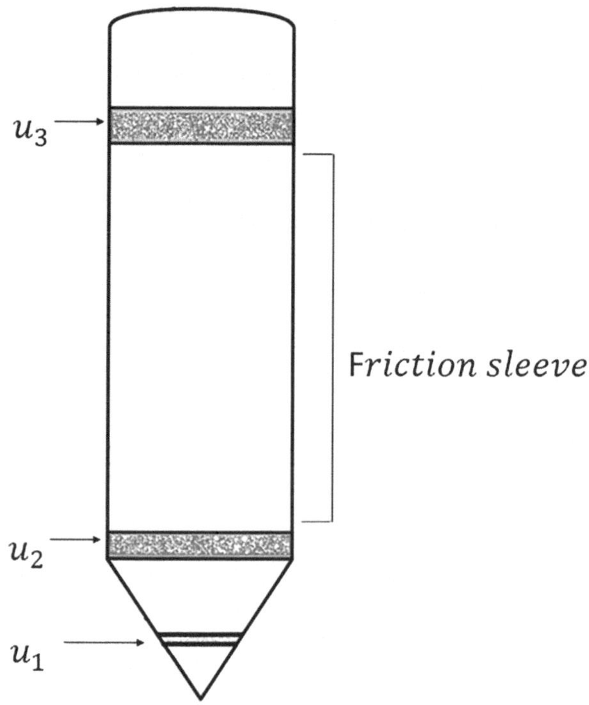

Fig. 9.1 Piezocone locations

Zhang et al. [27] presented a unique curve that can be linked operatively to the conventional coefficient by estimating the initial pore pressure at the shoulder of the piezocone in over-consolidated soils. Emetere and Alalade [28] proposed various concrete qualities for pile construction based on the results of CPT and geographical information systems (GIS). CPT and GIS helped to provide a spatial mapping of soil movements. Vardon and Peuchen [29] proposed a model based on CPT measurements to estimate the thermal properties of soils, including volumetric heat capacity and thermal conductivity, along continuous profiles. Doan and Lehane [30] investigated the relationship between CPT data and normally consolidated soils to identify soil behavior types and estimate the degree of over-consolidation and under-consolidation in unaged soils. Guan and Wang [31] studied the effect of spatial variability of soil properties on soil liquefaction potential and proposed a probabilistic model linking spatial variability to vertical liquefaction potential. Zhang et al. [32] improved the finite element method to correlate CPT measurements with soil characteristics.

9.2.2 *Undrained Shear Strength*

S_u represents its capacity to resist shear stress. This parameter is influenced by several factors, including the directional dependence of soil properties, deformation rate, mode of shear failure, and OCR. The selection of an appropriate S_u value depends on the specific design problem [33]. Laboratory tests, such as unconsolidated undrained triaxial tests (UUTs), and in situ tests, such as field vane tests (FVTs), are commonly used to determine soil strength properties [33, 34]. However, both laboratory and field test results may not fully capture the true soil characteristics. Factors such as device size, the volume of soil tested (particularly for interbedded or fissured soils), operator technique, shear kinematics, wall effects, device geometry, and soil disturbance can all influence the measured values [3]. There are various methods introduced to calculate S_u. Theoretical methods comprise: (1) cavity expansion theory; (2) classical bearing capacity theory; (3) analytical and numerical approaches using linear and nonlinear stress-strain relationships; (4) the principle of energy conservation combined with cavity expansion theory; (5) strain path theory [35]. Wang et al. [36] developed a theoretical model to predict S_u based on the liquidity index (LI) using critical state theory and the Mohr–Coulomb failure criterion. Ho and Hsieh [37] applied numerical analysis incorporating the plasticity index (I_p) for S_u prediction.

Laboratory testing provides another approach for determining S_u. Cheshomi et al. [6] established a correlation between the limit pressure and S_u obtained from uniaxial compression tests. Belkhatir et al. [7] proposed an equation linking the hydraulic conductivity of a soil blend comprising sand and silt with S_u values derived from undrained monotonic triaxial tests.

According to Robertson [34], in situ test results are generally more reliable for strength prediction because they more accurately reflect actual field stress conditions than laboratory tests. The relationship between S_u and CPT measurements has therefore received considerable attention. Theoretical principles provide a basis for interpreting CPT data, with bearing capacity theory—derived from Terzaghi's formulation—being widely used to relate q_c to S_u of cohesive soils. This relationship can be expressed as follows (Eq. 9.1):

$$S_u = \left(q_c - \sigma_{v0}\right) / N_k \tag{9.1}$$

N_k is the cone factor.

Theoretical approaches are the fundamental basis for developing empirical or semi-empirical equations. If the effect of u_2 is regarded, then (see Eq. 9.2):

$$S_u = \left(q_t - \sigma_{v0}\right) / N_k \tag{9.2}$$

These equations depend on the geometry of the instruments, soil properties, evaluation techniques, and other factors. Lunne and Kleven [38] demonstrated that N_k depends on I_p. They suggested I_p = 15. Various N_k have been proposed regarding diverse soil types and assessment methods. Moreover, some researchers considered the effective cone factor (N_{ke}) as presented in Eq. (9.3):

$$S_u = \left(q_t - \sigma_{v0}\right) / N_{ke} \tag{9.3}$$

Hong et al. [39] achieved N_{ke} ranging from 3 to 18 for Busan clay. Some other research suggested 1–10 and 6–12. Nevertheless, they pointed out that this kind of cone factor is not as appropriate as N_k and induces more error in the prediction of Busan clay. Cai et al. [35] proposed two values for the cone factor (equal to 15 or 10) for Lianyungang marine clay in Jiangsu province of China by analyzing CPT data, besides FVT data and previous research on this area.

The relationship between S_u and excess porewater pressure based on cavity expansion theory has received attention (Eq. 9.4).

$$S_u = \left(u_2 - u_0\right) / N_{\Delta u} \tag{9.4}$$

$N_{\Delta u}$ is the pore water pressure cone factor.

Depending on the soil type and situation, $N_{\Delta u}$ ranges from 2 to 20, but considering the consolidated-undrained triaxial compression test on soft soil, $N_{\Delta u}$ = 4–10 [40]. Li et al. [40] recommended a linear solution between S_u and excess porewater pressure. They determined the amount of pore pressure cone factor, which equals 8.8, at the Ningbo site.

Shear wave velocity (V_s) is another parameter to measure soil strength. In cohesive soils, σ'_{v0}, OCR and void ratio are influential parameters on V_s. Oh et al. [41] proposed an experimental correlation between S_u and V_s based on the relationship between the void ratio and σ'_{v0}.

Table 9.1 provides a summary of S_u correlations according to previous research, including the input parameters involved in each correlation model, the test site or the number of datasets analyzed, and the soil types involved in each study.

9.2.3 Machine Learning

ML is a field of computer science that enables computers to learn and comprehend data. It consists of algorithms that perform specific tasks and improve their performance through experience. The main advantage of ML is its practicality when dealing with problems that are too complex for traditional methods to handle. In other words, it supports data mining, which involves discovering patterns in large datasets that are not immediately apparent. All benefits of ML can be summed up in four phrases: (1) complex problems, including long lists of rules and regulations, can be simplified; (2) in fields in which traditional approaches are not fine; (3) fluctuating environments cause heterogeneous conditions incorporating a lot of variables; (4) when there are large amounts of data. MLs' categories are: (1) whether they are

Table 9.1 Summary of S_u correlations

Input parameters	Test site/number of datasets	Soil type	Reference
N_{ke}, q_t, u_2; $N_{ke} = 3–18$	Data from a site in Busan, Korea	Busan clay	[39]
q_t, σ_{v0}, N_{kt}; $N_{kt} = 15$ or 10	Data from a site in Lianyungang, China	Soft marine clay	[35]
V_s, e_r	Data from a site in Busan, Korea	Marine clay	[41]
$(u_2 - u_0)$, $N_{\Delta u}$; $N_{\Delta u} = 8.8$	Data from Ningbo test site	Clayey soils	[40]
q_{t-avg}, f_{s-avg}, σ_{v0}	70 test sites in Louisiana, US	Clayey soils	[42]
P_L	44 tests in Qom, Iran	Clayey soils	[6]
w, σ_v, LL, I_p, OCR, S_t	353 datasets	Clayey soils	[13]
q_c, f_s, OBW, PL, LL	304 datasets from Louisiana, US	Clayey soils	[43]
w, γ_d, LL, I_p, CI, e, g_s, q_p	111 datasets from the northern areas of the Nile River, Egypt	Sensitive alluvial Soft clayey soils	[4]
LI	103 samples	Clayey soils	[36]
SPT number, depth, and moisture content	328 data groups from Alberta, Canada	Clay till	[36]

taught with human control (supervised, unsupervised, reinforcement learning, and semi-supervised); (2) online and batch learning that are different based on immediate or gradual data learning; (3) instance-based or model-based learning that varies in learning and improvement methods. These methods can also be combined. Deep learning (DL) and ANNs enable computers to mimic the pattern recognition processes of the human brain. The origins of such ideas can be traced back to ancient philosophers. Plato and Aristotle were among the earliest thinkers who attempted to explain associationism—the idea that mental processes operate through associations between concepts and experiences. The theory of associationism was later refined through the work of several philosophers and psychologists. Thomas Hobbes (1588–1679) proposed that associations of sensations form simple experiences, which in turn combine into more complex ones. He also argued that coherence and frequency are the main factors maintaining associations. Hartley (2013) played a key role in popularizing associationism. He proposed that memory "vibrations" occur in the same brain areas as the primary sensory experiences. These vibrations form the basis of consciousness and contribute to the generation of complex ideas. This concept influenced the development of the Hebbian Learning Rule (HLR). Alexander Bain (1818–1903) contributed to the foundational ideas of the HLR. He demonstrated the connection between associative memory and the activity distribution of neural categories. Bain proposed a type of memory storage that could assemble required materials even in the presence of previously stored memories. He also explained the computational function that allows neural groupings to become active when multiple associations are to be stored. Bain further explored the construction of neural groupings and their relationship to memory, suggesting that the relevant perception of a neural grouping depends on its temporal characteristics. Subsequent research focused on the mechanisms underlying the strength or weakness of connections between neurons, attributed to changes in the intervening cell substance. Hebb [44] is widely regarded as the father of neural networks. In his seminal work,

The Organization of Behaviour, he introduced a learning approach that later became known as the HLR. However, certain limitations of his method became apparent, particularly its tendency toward instability: repeated co-occurrences could cause excessive increases in connection weights. Erkki Oja (1982) proposed a modification to the HLR to address this instability, demonstrating that his rule functions as an online update of the principal component analyzer. Although Hebb is credited with developing neural networks based on HLR, Warren McCulloch and Walter Pitts had modeled the first neural network 6 years prior, using electrical circuits to simulate neuron behavior. Their model is known as the MCP neural model. The perceptron, a fundamental concept in machine learning, was inspired by ideas from psychology.

The perceptron is an early ANN model, drawing on studies of animal and human learning. It served as a foundational building block for more complex neural network architectures used in modern ANNs. In 1962, Novikoff [45] established the convergence theorem for the perceptron algorithm, proving that it would converge in a finite number of steps under certain conditions. By 1960, there were already several questions about the proper methods for testing and evaluating ANN models. Kozinets [46] studied a simple converging algorithm that tends toward an optimal supporting hyperplane. MDM, another algorithm, was proposed by Mitchell et al. [47], which employs non-smooth optimization methods. Yakubovich [48, 49] refined a principled approach to mathematical pattern recognition theory and called it the "method of recurrent goal inequalities." This method constructs a system of inequalities capable of providing solutions for an infinite number of them, leading to improvements in dynamical systems. All these modification attempts induced (1) signal isolation from noise description and analyses of scenes, (2) handwriting recognition training, and (3) aerial photographs [50]. Bregman [51] proposed a solution to a problem related to city planning. He introduced a repeated algorithm for locating the intersection of convex sets in a Hilbert space. The following year, he extended these results to establish a general framework for convex optimization [51]. Additionally, Bregman proposed a method linking his approach to Lyapunov function convergence proofs. This breakthrough was used in several areas, like ML, image deblurring, clustering, data reconstruction, image segmentation, etc. In addition, there are several convergent algorithms in a finite number of legs, as well as applications of Chebyshev estimation and optimal control. The mentioned algorithms are deterministic methods that cover uncertainty parameters as elements of bounded sets. Statistical rules of communications and decision theory constitute the base of statistical systems [50]. Tsypkin [52, 53] brilliantly enhanced a method based on average risk minimization, which includes a significant number of algorithms suggested by different authors as particular cases. Although Feldbaum [54] was the first to propose the concept of average risk minimization for the control of uncertain systems, Tsypkin was the first to incorporate stochastic methods into this approach. The idea of stochastic approximation was obtained from mathematical statistics. Tsypkin's method has been a standard scientific work in the study of adaptation and learning algorithms. The average of the cost function is the root of Tsypkin's performance in adaptation and learning problems [48]. The cost function term has caused the generation of a lot of ML algorithms. Convexity and mathematical components under the Robbins-Monro conditions make the stochastic approximation capable of establishing convergence of the algorithms. Some researchers published a book about solving the limitations of complex problems by perceptrons. They said that perceptrons are not able to encompass some logical functions like XOR or NXOR. These restrictions caused AI research to lose financial support for over two decades. This time is called the "winter of AI." Nonetheless, this did not cause complex algorithms to lose their fans. Between 1970 and 1980, further research was conducted on the learning capabilities of multilayer NNs. In 1980, a multilayered NN algorithm was studied by Fukushima [55]. It is named "neocognitron." The result was the foundation of the backpropagation learning algorithm in the mid-1980s. However, its primary idea was developed in the early 1960s. Rumelhart et al. [56] proposed a groundbreaking method, "backpropagation," in their article, but some disadvantages of this method, especially when data cannot be classified separately, became clear [57, 58]. Therefore, investments in ML diminished again in the 1990s. This may be called "the second winter of AI." Another breakthrough was made when Cortes and Vapnik [58] proposed a plan for nonseparable issues by creating a novel version of the SVM algorithm under the name "support-vector networks." At the beginning of the twenty-first century, some trends caused impressive developments in ML. (1) Scientists generated new algorithms for a large quantity of data; (2) a reduction in the cost of parallel computing and memory. In more detail, Google wanted Hadoop and MapReduce technology together to distribute huge amounts of data amongst ordinary processes. At this time, GPU proved itself as a monopolist in the ML area; (3) RAMs became cheaper; thus, using large amounts of data in memory was possible; (4) the Apache Spark framework was invented for distributing unstructured or weakly structured data. The growth in multilayer MMs led to deep neural networks (DNNs) which Dechter [59] is famous for the idea of this new algorithm in 1986.

9.2.4 Machine Learning and Civil Engineering

Some studies are trying to find geotechnical relationships based on CPT values and ML. Zhao et al. [60] invented a method to determine efficient points of a site for soundings by CPT to know the spatial variability with the help of information entropy and Bayesian compressive sensing. Kennedy et al. [61] studied a method to find the efficient length of in situ piles for towers with NNs and limited soil information acquired by a seismic cone penetrometer. Ozsagir et al. [62] made a comparison between various kinds of ML models to discover the soil liquefaction potential by the standard penetration test. They identified a decision tree model to suit the goals best. A probabilistic time-dependent relationship based on MLAs for stability evaluation of reservoir slopes was suggested [36]. Probability summation method as an ANN to estimate the probability of slope failure [63]. Kazemi et al. [64] used MLAs to illustrate curves of seismic steel moment-resisting frame capacity. Cheng et al.

[65] improved a DNN to consider ground motions and critical accelerations for seismic stability of slopes. Wang et al. [36] investigated a hybrid modeling method to consider the effect of more soil uncertainties on embankment settlement for the optimization of basal reinforced piled embankments. He et al. [66] considered spatial variability when calculating the factor of safety of slopes using MLAs (the NNs and the support-vector regression) to reduce time consumption significantly. He et al. [67] suggested modifications in ANN algorithms to produce surrogate models with variable parameters instead of deterministic ones. He et al. [68] evaluated the idea of connecting advanced models in geotechnics in practice regarding special variability, which is capable of covering all required materials and boundary conditions by being trained on a large number of inputs. He et al. [68] also contributed to easing the geotechnical computations by considering any soil properties, load conditions, or spatial variabilities. Therefore, their model minimizes or makes unnecessary the requirements for numerical simulations. Yang et al. [69] exploited a combination of a Bayesian model, a random forest and a CNN to classify the rock mass quality for excavation by TBM. Zhang et al. [27] incorporated spatial variability into model training to determine soil pollution levels by MLAs. Wang et al. [36] used the VNIR technique to obtain soil pollution indexes and find an appropriate relationship through MLAs. Zhang et al. [16] collected discussions, insights, and challenges about using ML in geoscience and geotechnical engineering.

9.3 Hyperparameter Optimization

Unlike weights and biases, hyperparameters cannot be updated directly during training. However, hyperparameter optimization (HPO) enables their systematic adjustment, as shown in Eq. (9.5)

$$x^{*} = \arg\min_{x \in X} f(x) \tag{9.5}$$

Here, $f(x)$ denotes the objective function, which represents the error rate. x^{*} is the set of optimal hyperparameters within the search space X (all possible configurations), obtained by minimizing $f(x)$. Thus, HPO aims to identify hyperparameter settings that minimize the error rate on a validation set.

9.4 Bayesian Deep Neural Network

In this study, deep learning (DL) is combined with variational inference (VI), a Bayesian inference method, to account for uncertainty in the weights and biases. This integration enables the models to quantify the uncertainty associated with these parameters. According to Bayes' theorem, the prior and posterior distributions are expressed in Eqs. (9.6) and (9.7) as representations of uncertainty:

$$P(W|D) = \frac{P(W) \cdot P(D|W)}{P(D)} \quad (9.6)$$

$$P(D) = \int P(D|W) P(W) dW \quad (9.7)$$

where W represents the weights and biases and D denotes the training dataset. $P(D)$ is the normalizing constant (or model evidence), $P(W)$ is the prior, and $P(W|D)$ is the posterior distribution.

$$P(y_{new}|x_{new}, D) = \int_W P(y_{new}|x_{new}, D, W) P(W|D) dW \quad (9.8)$$

Equation (9.8) represents the likelihood of new data, which plays a key role in determining the updated length of the confidence interval, making it either narrower or broader. The prior distribution reflects an initial assumption about the unknown parameter before any data are observed. This parameter is treated as a random variable with its own probability distribution. Since there is initially no clear knowledge of a suitable prior, it is commonly assumed to follow a normal distribution, as shown in Eq. (9.9) [4].

$$P(W) = N(0, \lambda I) \quad (9.9)$$

9.5 Data Description

This study employs the CLAY/10/7490 dataset from http://140.112.12.21/issmge/tc304.htm, which provides 10 clay parameters and covers 30 districts worldwide. The database consists of insensitive to quick clays, with OCR values ranging from 1 to 86. After preprocessing, the 349-entry dataset was used to train the DL model. The dataset includes nine input variables: σ'_{v0}, σ'_p, *OCR*, *LL*, *PL*, *w*, q_t, u_0, u_2 and one output variable: S_u collected from 11 countries (including the USA, Sweden, Canada, the UK, Hong Kong, Singapore, Malaysia, Brazil, Norway, Italy, and Venezuela) with OCR between 1 and 26. A descriptive statistics table of the data is presented in Table 9.2, showing the statistical characteristics of the datasets, which were rescaled using *Z*-score normalization to remove scale effects. All variables are expressed in kPa, except for *OCR*, *LL*, *PL*, and *w*, which are dimensionless. Notably, q_t and u_2 exhibit the highest variability.

Table 9.2 Descriptive statistics of the dataset

Statistics	σ'_{v0}	σ'_{p}	***OCR***	**LL**	**PL**
Min	6.393	20.000	0.808	23.338	2.732
Max	506.234	1522.950	26.732	236.395	68.522
Mean	93.214	188.756	2.429	67.262	27.253
Std. deviation	78.088	214.893	2.762	25.409	9.154
25th percentile	37.948	64.009	1.171	49.522	20.744
50th percentile	67.495	121.429	1.578	67.063	25.632
75th percentile	131.530	248.320	2.440	80.957	32.148
Statistics	q_t	w	u_0	u_2	s_u
Min	175.338	15.546	0.000	22.333	5.559
Max	4264.531	170.485	778.197	2937.448	219.226
Mean	921.894	61.284	143.747	550.797	34.701
Std. deviation	816.700	25.675	149.158	527.603	34.414
25th percentile	356.892	43.816	50.783	210.956	12.979
50th percentile	624.592	62.295	90.866	354.937	23.813
75th percentile	1125.584	75.852	168.242	639.405	39.459

9.6 Data Split

Firstly, 60% of the complete dataset was used for training, 20% for testing, and the remaining 20% for validation. S_u is the output, and the other variables are the inputs.

9.7 Feature Scaling

To reduce the impact of noisy data, potential outliers, and differences in scale, *Z*-score standardization has been applied to improve the performance of the DL models. Consequently, the rescaled dataset has a mean of 0 and a standard deviation of 1 (Eq. 9.10).

$$Z - \text{score} = \frac{x - \mu}{q} \tag{9.10}$$

where x is a datapoint, μ is mean, and q is the standard deviation.

This technique is also very useful for BDNN due to its effects on the magnitude of the posterior distribution of weights and biases [17]. In a comparison between *Z*-score standardization and min-max scaling after conducting trials, *Z*-score standardization impacts the accuracy of the BDNN model.

9.8 Bayesian Deep Neural Network with Nine Inputs

In this study, the BDNN model, in addition to VI, incorporates KL divergence, learning rate tuning, activation functions, and a dropout regularization technique (dropout rate = 0.2) to prevent overfitting and model nonlinear relationships. Furthermore, Bayesian optimization (BO) uses a validation dataset to evaluate the performance of 400 DNN configurations based on mean squared error (MSE) and select the best model.

The model has nine input variables and one output variable. The output layer consists of a single neuron, with no activation function applied. Other hyperparameters are optimized using BO. The optimal ranges of the four hyperparameters used to determine the best model configuration are presented in Table 9.3.

To identify the most optimal values, BO uses root mean squared error (RMSE) as the objective metric to minimize and determine the best model architecture. Table 9.4 presents the optimal BDNN configuration based on the specified hyperparameter ranges and additional parameters. It should be noted that the number of neurons in the output layer, the activation function in the output layer, the dropout rate, and the learning rate were determined manually through trial and error.

Table 9.5 presents the performance metrics for the training-validation and testing datasets. In most cases, the testing dataset shows better performance, except for MSE, which is lower in the training-validation set. This discrepancy may result from the effect of optimized hyperparameters or the sensitivity of VI to noisy data, which may be less prevalent in the testing dataset.

Table 9.3 The hyperparameters of the BDNN

Hyperparameters	Optimization range
The number of hidden layers	(1, 3)
The number of neurons	(1, 90)
Activation functions	(Tanh; Sigmoid; ReLU)
Learning rate	(0.0001; 0.001; 0.01)

Table 9.4 The BDNN architecture with nine inputs

Hyperparameters	Value
The number of neurons in the first dense layer	57
The number of neurons in the second dense layer	29
The number of neurons in the output layer	1
The number of hidden layers	2
The number of input layers	1
The number of output layers	1
The activation function in the first dense layer	Tanh
The activation function in the second dense layer	ReLU
The activation function in the output layer	Linear (no activation function)
The number of epochs	86
Learning rate	0.01
Dropout rate	0.2

The correlations between the measured S_u and the predicted ones by BDNN with nine inputs for the training-validation and testing datasets are illustrated in Fig. 9.2.

Table 9.5 The evaluation metrics in the BDNN model with nine inputs

Evaluation metrics	Training-validation set	Testing set
MAE	9.888	9.691
MSE	192.955	221.350
R-squared score	0.814	0.869
VAF (%)	81.418	87.325
a20_index	0.369	0.414

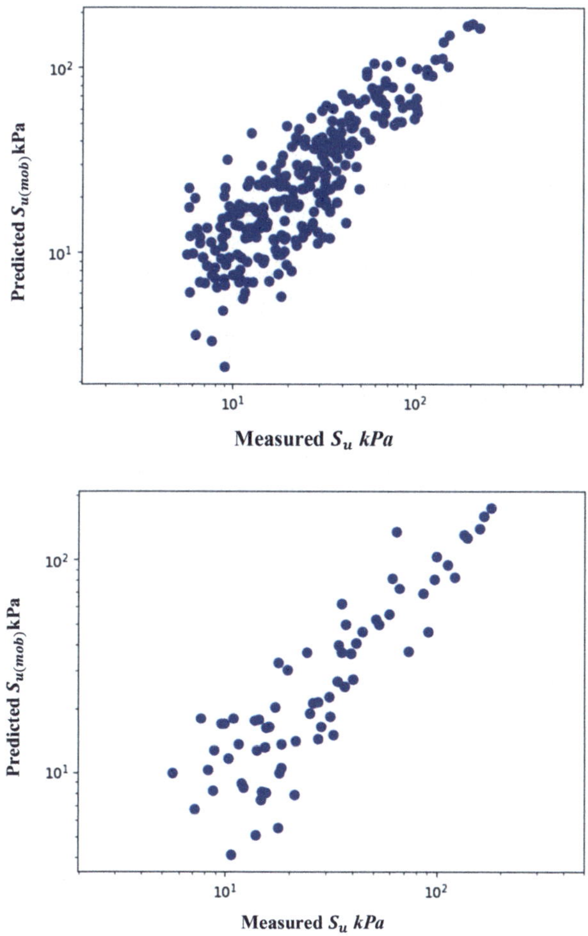

Fig. 9.2 Predicted values versus measured ones by the BDNN with nine inputs

9.9 Pearson Cross-Correlation

Pearson Cross-Correlation (PCC) computes the correlation coefficient value considering a linear relationship between two variables. PCC does not regard time shifts while calculating the correlation between two variables. The formula for PCC is given below:

$$PC(X,Y) = \frac{n(\Sigma XY) - (\Sigma X)(\Sigma Y)}{\sqrt{\left[n\Sigma X^2 - (\Sigma X)^2\right]\left[n\Sigma Y^2 - (\Sigma Y)^2\right]}} \tag{9.11}$$

n is the number of data points, and X and Y represent the individual data points in each dataset.

9.10 Correlation Analysis

The variation in correlation between the input parameters and S_u is demonstrated in Table 9.6 below. Pearson's correlation coefficient was applied to determine the correlations between each pairwise variable. According to Table 9.6, the correlations between q_t, σ'_p, u_2, σ'_{v0}, and S_u are the strongest and relatively positive.

Considering the selected four input variables (q_t, σ'_p, u_2, and σ'_{v0}) and one output (S_u) variable, 431 datasets from 11 countries were analyzed: the USA, Sweden, Canada, the UK, Hong Kong, Singapore, Malaysia, Brazil, Norway, Italy, and Venezuela. Table 9.7 lists summary statistics. Notably, q_t and u_2 exhibit the highest variability.

Table 9.6 Pearson's correlation matrix

	σ'_{v0}	σ'_p	OCR	LL	PL	w	q_t	u_0	u_2	S_u
σ'_{v0}	1									
σ'_p	0.471	1								
OCR	−0.091	0.430	1							
LL	−0.108	−0.127	−0.072	1						
PL	−0.106	−0.125	−0.063	0.792	1					
w	−0.242	−0.416	−0.237	0.896	0.769	1				
q_t	0.573	0.866	0.540	−0.209	−0.204	−0.538	1			
u_0	0.725	0.152	−0.105	−0.017	−0.005	−0.286	0.413	1		
u_2	0.682	0.741	0.141	−0.161	−0.169	−0.502	0.898	0.559	1	
S_u	0.689	0.734	0.150	−0.084	−0.054	−0.194	0.754	0.190	0.691	1

Table 9.7 Descriptive statistics of the dataset

Statistics	σ'_{v0}	σ'_p	q_t	u_2	s_u
Min	6.393	18.723	175.338	22.333	4.595
Max	506.234	1522.950	4264.531	2937.448	219.226
Mean	95.989	197.326	969.854	556.355	37.773
Std. deviation	76.637	203.335	806.287	504.109	35.274
25th percentile	39.744	69.796	384.641	213.793	14.382
50th percentile	74.488	137.000	740.766	393.328	27.680
75th percentile	134.366	265.887	1284.164	665.011	44.132

Table 9.8 The BDNN architecture with four inputs

Hyperparameters	Value
The number of neurons in the first dense layer	53
The number of neurons in the second dense layer	45
The number of neurons in the output layer	1
The number of hidden layers	2
The number of input layers	1
The number of output layers	1
The activation function in the first dense layer	ReLU
The activation function in the second dense layer	Tanh
The activation function in the output layer	Linear (no activation function)
The number of epochs	108
Learning rate	0.01
Dropout rate	0.2

9.11 Bayesian Deep Neural Network with Four Inputs

After selecting the four input variables and compiling a dataset of 431 records, the model's HPs were optimized using 20% of the data, trained on 60%, and tested on the remaining 20%.

Considering the optimal hyperparameter ranges in Table 9.3 and excluding eight outliers with large prediction errors, Table 9.8 presents the final BDNN configuration based on these ranges and additional parameters. The number of neurons in the output layer, the activation function in the output layer, the dropout rate, and the learning rate were determined manually through trial and error.

Table 9.9 presents the MAE, MSE, R-squared score, VAF (%), and a20_index for both the training-validation and testing sets. Overall, the training-validation set outperforms the testing set across all metrics except for the a20_index.

The correlations between the measured S_u and the predicted ones by BDNN with four inputs for the training-validation and testing datasets are illustrated in Fig. 9.3.

Table 9.9 The evaluation metrics in the BDNN model with four inputs

Evaluation metrics	Training	Testing set
MAE	10.948	9.189
MSE	313.821	206.502
R-squared score	0.743	0.766
VAF (%)	74.650	76.923
a20_index	0.396	0.412

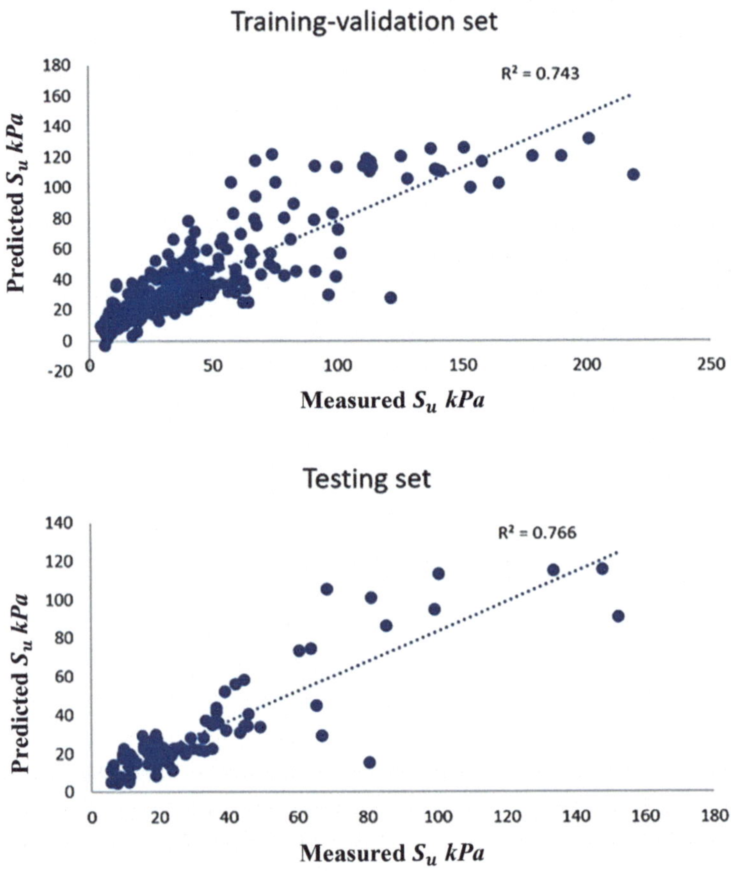

Fig. 9.3 Predicted values versus measured ones by the BDNN with four inputs

9.12 Conclusion

This research demonstrates the effectiveness of the BDNN model in predicting the undrained shear strength of clayey soils based on correlations with CPTU data. The main conclusions of this research are:

- *Z*-score standardization affects the accuracy of the BDNN model. This highlights the sensitivity of DL algorithms to data scaling.
- In this study, eight datasets were identified as outliers. The findings indicate that removing these outliers improved the performance of BDNN models, resulting in approximately a 24% increase in R^2.
- The performance of conventional DL modeling increases when incorporating Bayesian optimization. This ensures more optimal and reliable predictions.
- The BDNN demonstrates effective generalization, outperforming on the test set likely due to its ability to quantify uncertainty and avoid overfitting to noise in the training data.
- Despite the reduction in the number of input variables, the difference in performance between the two models is not statistically significant. Several metrics for the four-input model are slightly better than for the nine-input model, indicating that the BDNN with VI can provide reliable predictions even with fewer inputs.
- In our BDNN, performance on the testing set exceeds that on the training-validation set, suggesting that Bayesian regularization via variational inference helps the model avoid overfitting noisy training examples and generalize well to unseen data.
- Including CPTU measurements in the input set improved the accuracy of DL-based predictions of S_u, especially for models with a reduced number of inputs.

AI Use Disclosure Statement The authors used generative AI technologies only for proofreading. This includes minor corrections to grammar, spelling, and syntax. No AI was used for ideation, data analysis, or writing original content.

Declaration The authors used generative AI technologies only for proofreading. This includes minor corrections to grammar, spelling, and syntax.

References

1. Elkateb T, Chalaturnyk R, Robertson PK (2003) An overview of soil heterogeneity: quantification and implications on geotechnical field problems. Can Geotech J 40(1):1–15
2. Phoon KK, Kulhawy FH (1999) Characterization of geotechnical variability. Can Geotech J 36(4):612–624
3. Stefanow D, Dudziński PA (2021) Soil shear strength determination methods—state of the art. Soil Tillage Res 208:104881
4. Elsawy MB, Alsharekh MF, Shaban M (2022) Modeling undrained shear strength of sensitive alluvial soft clay using machine learning approach. Appl Sci 12(19):10177
5. Rémai Z (2013) Correlation of undrained shear strength and CPT resistance. Period Polytech Civil Eng 57(1):39–44
6. Cheshomi A, Bakhtiyari E, Khabbaz H (2020) A comparison between undrained shear strength of clayey soils acquired by "PMT" and laboratory tests. Arab J Geosci 13:1–13
7. Belkhatir M, Schanz T, Arab A (2013) Effect of fines content and void ratio on the saturated hydraulic conductivity and undrained shear strength of sand–silt mixtures. Environ Earth Sci 70:2469–2479

8. Der Kiureghian A, Ditlevsen O (2007) Epistemic or aleatory? Does it matter? In: Proceedings of the special workshop on risk acceptance and risk communication. Stanford University, Stanford, CA. http://www.ripid.ethz.ch/Paper/DerKiureghian_paper.pdf
9. Lunne T, Powell JJ, Robertson PK (2002) Cone penetration testing in geotechnical practice. CRC Press
10. Lunne T, Berre T, Strandvik S (1997) Sample disturbance effects in soft low plastic Norwegian clay. In: Symposium on recent developments in soil and pavement mechanics. CAPES-Fundacao Coordenacao do Aperfeicoamento de Pessoal de Nivel Superior; CNPq-Conselho Nacional de Desenvolvimento Cientifico a Tecnologico; FAPERJ-Fundacao de Ampora a Pesquisa do Estado do Rio de Janeiro; FINEP-Financiadora de Estudos e Projetos
11. Sowers GF (1979) Introductory soil mechanics & foundations. Geotech Eng 92:114–117
12. Clayton CRI, Matthews MC, Simons NE (1995) Site investigation, vol 436. Blackwell Science, Oxford, p 441
13. Ching J, & Phoon KK (2014) Transformations and correlations among some clay parameters—the global database. Canadian Geotechnical Journal 51(6):663–685
14. Zhang W, Wu C, Zhong H, Li Y, Wang L (2021) Prediction of undrained shear strength using extreme gradient boosting and random forest based on Bayesian optimization. Geosci Front 12(1):469–477
15. Chongzhi W, Lin W, Zhang W (2021) Assessment of undrained shear strength using ensemble learning based on Bayesian hyperparameter optimization. In: Modeling in geotechnical engineering. Academic Press, pp 309–326
16. Zhang W, Gu X, Tang L, Yin Y, Liu D, Zhang Y (2022) Application of machine learning, deep learning and optimization algorithms in geoengineering and geoscience: comprehensive review and future challenge. Gondwana Res 109:1–17
17. Oleksiienko I, Tran DT, Iosifidis A (2023) Variational neural networks. Procedia Comput Sci 222:104–113
18. Zhou J, Koopialipoor M, Li E, Armaghani DJ (2020) Prediction of rockburst risk in underground projects developing a neuro-bee intelligent system. Bull Eng Geol Environ 79(8):4265–4279
19. Barkhordari M, Armaghani D, Asteris P (2023) Structural damage identification using ensemble deep convolutional neural network models. Comput Model Eng Sci 134(2):835
20. Barkhordari MS, Armaghani DJ, Mohammed AS, Ulrikh DV (2022) Data-driven compressive strength prediction of fly ash concrete using ensemble learner algorithms. Buildings 12(2):132
21. Zhou J, Guo H, Koopialipoor M, Jahed Armaghani D, Tahir MM (2021) Investigating the effective parameters on the risk levels of rockburst phenomena by developing a hybrid heuristic algorithm. Eng Comput 37(3):1679–1694
22. Armaghani DJ, Mohamad ET, Narayanasamy MS, Narita N, Yagiz S (2017) Development of hybrid intelligent models for predicting TBM penetration rate in hard rock condition. Tunn Undergr Space Technol 63:29–43
23. Asteris PG, Armaghani DJ (2025) An empirical-driven machine learning (EDML) approach to predict PPV caused by quarry blasting. Bull Eng Geol Environ 84(4):200
24. Afrazi M, Jahed Armaghani D, Afrazi H, Fattahi H, Samui P (2025) Real-time monitoring of tunnel structures using digital twin and artificial intelligence: a short overview. Deep Undergr Sci Eng. https://doi.org/10.1002/dug2.70029
25. Armaghani DJ, Rasekh H, Asteris PG (2024) An advanced machine learning technique to predict compressive strength of green concrete incorporating waste foundry sand. Comput Concr 33(1):77–90
26. Kang C, Brotherton M, Anderson K, Semeniuk K, Mayer G, Law A et al (2024) Examination of the correlation between SPT and undrained shear strength: case study of clay till in Alberta, Canada. Eng Geol 334:107510
27. Zhang W, Liu K, Wang D, Zheng J (2023) Coefficient of consolidation measured by cone penetration tests in overconsolidated cohesive soils. Ocean Eng 276:114301
28. Emetere ME, Alalade GM (2023) Spatial analysis of cone penetration test measurement within stadium site. Pract Period Struct Des Constr 28(2):04023005

29. Vardon PJ, Peuchen J (2021) CPT correlations for thermal properties of soils. Acta Geotech 16(2):635–646
30. Doan LV, Lehane BM (2021) CPT data in normally consolidated soils. Acta Geotech 16(9):2877–2885
31. Guan Z, Wang Y (2022) CPT-based probabilistic liquefaction assessment considering soil spatial variability, interpolation uncertainty and model uncertainty. Comput Geotech 141:104504
32. Zhang W, Zou JQ, Zhang XW, Yuan WH, Wu W (2021) Interpretation of cone penetration test in clay with smoothed particle finite element method. Acta Geotech 16(8):2593–2607
33. Mayne PW (2007) Cone penetration testing, vol 368. Transportation Research Board
34. Robertson PK (2009) Interpretation of cone penetration tests—a unified approach. Can Geotech J 46(11):1337–1355
35. Cai GJ, Liu SY, Tong LY, Du GY (2010) Field evaluation of undrained shear strength from piezocone penetration tests in soft marine clay. Mar Georesour Geotechnol 28(2):143–153
36. Wang Q, Qiu S, Zheng H, Zhang R (2023) Undrained shear strength prediction of clays using liquidity index. Acta Geotech 19:1–18
37. Ho IH, Hsieh CC (2013) Numerical modeling for undrained shear strength of clays subjected to different plasticity indexes. J Geoeng 8(3):91–100
38. Lunne T, Kleven A (1981) Role of CPT in North Sea foundation engineering. In: Cone penetration testing and experience. ASCE, pp 76–107
39. Hong SJ, Lee MJ, Kim JJ, Lee WJ (2010) Evaluation of undrained shear strength of Busan clay using CPT. In: Proceedings of the 2nd international symposium on cone penetration testing, CPT, vol 10
40. Li X, Cai G, Liu S (2018) Evaluation of undrained shear strength of clay using the CPTU pore pressure method. Soil Mech Found Eng 55(3):162–167
41. Oh TM, Bang ES, Cho GC, Park ES (2017) Estimation of undrained shear strength for saturated clay using shear wave velocity. Mar Georesour Geotechnol 35(2):236–244
42. Abu-Farsakh MY, Mojumder MAH (2020) Exploring artificial neural network to evaluate the undrained shear strength of soil from cone penetration test data. Transp Res Rec 2674(4):11–22
43. Chen H, Ya G, Li J, Cui H, Yao X (2022) Predicting undrained shear strength of soil from cone penetration test data applying optimized RBF approaches. J Appl Sci Eng 26(1):121–130
44. Didier JP, Bigand E (2011) Rethinking physical and rehabilitation medicine: new technologies induce new learning strategies. Springer Science & Business Media
45. Novikoff AB (1962) On convergence proofs on perceptrons. In: Proceedings of the symposium on the mathematical theory of automata, vol 12(1), pp 615–622
46. Kozinets BN (1964) On one algorithm for learning a linear perceptron. In: Vichislitelnaia Tekhnika i Voprosi Programirovania, vol 3
47. Mitchell BF, Dem'yanov VF, Malozemov VN (1974) Finding the point of a polyhedron closest to the origin. SIAM J Control 12(1):19–26
48. Yakubovich VA (1965) Certain general theoretical principles in the design of learning pattern recognition systems, part I. Vichislitelnaia Tekhnika i Voprosi Programirovania 4:3–71
49. Yakubovich VA (1966) Recurrent finitely convergent algorithms for solving systems of inequalities. In: Doklady akademii nauk, vol 166(6). Russian Academy of Sciences, pp 1308–1311
50. Fradkov AL (2020) Early history of machine learning. IFAC PapersOnLine 53(2):1385–1390
51. Bregman LM (1967) The relaxation method of finding the common point of convex sets and its application to the solution of problems in convex programming. USSR Comput Math Math Phys 7(3):200–217
52. Tsypkin YZ, Nikolic ZJ (1971) Adaptation and learning in automatic systems, vol 73. Academic Press, New York
53. Tsypkin I (1966) Adaptation, training and self-organization in automatic systems (Adaptation, learning and self-learning in automatic systems, discussing stochastic approximation, pattern recognition, adaptive filters, etc.). Autom Remote Control 27:16–51
54. Feldbaum AA (1960) Dual control theory, automation and remote control, vol 21, Parts I, II, April 1961, pp 874–880; May 1961, pp 1033–1039

55. Fukushima K (1980) Neocognitron: a self-organizing neural network model for a mechanism of pattern recognition unaffected by shift in position. Biol Cybern 36(4):193–202
56. Rumelhart DE, Hinton GE, Williams RJ (1986) Learning internal representations by error propagation. In: Rumelhart DE, McClelland J (eds) Parallel distributed processing, explorations in the microstructure of cognition, vol 1. MIT Press
57. Brady ML, Raghavan R, Slawny J (1989) Back propagation fails to separate where perceptrons succeed. IEEE Trans Circuits Syst 36(5):665–674
58. Cortes C, Vapnik V (1995) Support-vector networks. Mach Learning 20:273–297
59. Dechter R (1986) Learning while searching in constraint-satisfaction problems
60. Zhao T, Wang Y, Xu L (2021) Efficient CPT locations for characterizing spatial variability of soil properties within a multilayer vertical cross-section using information entropy and Bayesian compressive sensing. Comput Geotech 137:104260
61. Kennedy DV, Guzina BB, Labuz JF (2023) A machine learning framework for in situ sensing of pile length from seismic cone penetrometer data. Comput Geotech 159:105505
62. Ozsagir M, Erden C, Bol E, Sert S, Özocak A (2022) Machine learning approaches for prediction of fine-grained soils liquefaction. Comput Geotech 152:105014
63. Aminpour M, Alaie R, Khosravi S, Kardani N, Moridpour S, Nazem M (2023) Slope stability machine learning predictions on spatially variable random fields with and without factor of safety calculations. Comput Geotech 153:105094
64. Kazemi F, Jankowski R (2023) Machine learning-based prediction of seismic limit-state capacity of steel moment-resisting frames considering soil-structure interaction. Comput Struct 274:106886
65. Cheng Y, Wang J, He Y (2023) Prediction models of newmark sliding displacement of slopes using deep neural network and mixed-effect regression. Comput Geotech 156:105264
66. He X, Xu H, Sabetamal H, Sheng D (2020) Machine learning aided stochastic reliability analysis of spatially variable slopes. Comput Geotech 126:103711
67. He X, Wang F, Li W, Sheng D (2021) Efficient reliability analysis considering uncertainty in random field parameters: trained neural networks as surrogate models. Comput Geotech 136:104212
68. He X, Xu H, Sheng D (2022) Ready-to-use deep-learning surrogate models for problems with spatially variable inputs and outputs. Acta Geotech 18:1–18
69. Yang W, Zhao J, Li J, Chen Z (2023) Probabilistic machine learning approach to predict incompetent rock masses in TBM construction. Acta Geotech 18:1–19

Chapter 10
Sustainable Data-Driven Metaheuristic Models for Strength Evaluation of Marine Clay Treated with Recycled Tiles

Mahzad Esmaeili-Falak

Abstract As a final objective of several geotechnical experimental trials, a dependable assessment of the cohesive soils specification in conjunction with recycled materials was considered. This work utilized many attributes, including critical physical and mechanical properties of marine clay (*MC*) that were altered with recycled tiles (*RT*), to determine the unconfined compressive strength (*UCS*). Multiple innovative hybrid approaches were developed, integrating the least square support vector regression (*LSSVR*) analysis with different optimization algorithms like the golden jackal algorithm (*GJA*), artificial rabbit optimizer (*ARA*), and dwarf mongoose algorithm (*DMA*). Optimization techniques have been applied to the *LSSVR* frameworks to find the optimal hyperparameter value (*LSGJA*, *LSARA*, and *LSDMA*). The results indicate that the *LSGJA*, *LSARA*, and *LSDMA* methods have significant potential to accurately forecast the *UCS*. During the learning and examination stages, the R^2 and indices of agreement (*IA*) values for the *LSGJA* approach were found to be, respectively, 0.992 and 0.9962 and 0.998 and 0.999. The sensitivity analysis depicted that the simulation of *UCS* was highly affected by all attributes higher than 0.8435 (which was related to curing time (*CT*)), where the highest impact belonged to maximum dry density (*MDD*) at 0.9594, followed by *PH* at 0.9524.

Keyword Metaheuristics · *LSSVR* · Marine clay · Recycled tiles · Unconfined compressive strength

M. Esmaeili-Falak (✉)
Department of Civil Engineering, NT.C., Islamic Azad University, Tehran, Iran
e-mail: mahzad.ef@iau.ac.ir

M. Mirrashid et al. (eds.), *Intelligent Predictive Systems*, Emerging Trends in Mechatronics, https://doi.org/10.1007/978-981-95-7766-8_10

10.1 Introduction

The rapid expansion of both the economy and population in developing and affluent countries has depleted natural resources. Because of funding and land constraints, researchers, engineers, and land planners have to find ways to restore soft coastal earth. Most soft soils encountered by roads, railroads, and pavement projects must be rehabilitated. Building on soft soils is rare in tropical locations despite its increased need. Many old and contemporary solutions for stabilizing unstable soils were applied to satisfy building requirements. Soil alteration alters physical and technical properties to meet performance objectives. Soil alteration uses several methods and inputs that change soil properties [1]. Compaction [2], chemical additions [3], biological use of bacteria [4], electrical current injection, and a combination of these methods can all be employed to improve soft soil. Chemical additives appear to be the most popular due to their ability to rapidly enhance soil's mechanical specifications. Soil additives may be classified into conventional and non-conventional categories [5]. Chemical admixtures, fly-ash, and bituminous materials can be used as conventional and non-traditional soil amendments [6, 7]. Unconventional additives include enzymes, silicates, waste materials, and acids [8].

Utilizing waste materials in soft soil restoration has been promoted to decrease environmental issues and increase sustainability. Waste products were used since chemical soil stabilization was expensive. Multiple authors have raised concerns concerning geotechnical design using waste materials [9, 10]. Chemical additions reduce the need for costly chemical stabilizers in solid waste. It is occasionally applied to enhance the mechanical specifications of clayey soils due to waste products' environmental impact [11, 12]. Utilizing construction waste to amend soft soil will reduce waste and save natural resources. Waste generated by ceramic tile firms and building sites was repurposed by incorporating it into paper to strengthen unstable clay. About 7–30% of the overall ceramic tile production is considered discarded material [13, 14]. Water is introduced at various stages of the manufacturing process and generated during tile cutting, leading to the formation of a discarded material product with a mud-like consistency. This discarded material is collected in a slurry and dumped outside the building into the environment around it. There are a lot of fine materials in this muck, and the fine elements in the trash will be discharged into the air after the water has dried entirely. Ceramic tile trash from construction sites is regularly dumped in landfills, legally or illegally. This waste may pollute soil, take up space, and harm plants [15]. Soft soils are employed in geotechnical planning for slope stabilization, subgrade material for pavements, and foundation support for structures. Soft clay functions badly during building because it is inherently weak [16]. It is often weak and has poor properties due to the high content of organic materials and moisture. Moreover, low permeability and operational uncertainty have been connected to it [17]. It is often the case that the liquid limit of soft clay is less than its in situ water content [18]. The dominant minerals in clayey soils are kaolinite, illite, and montmorillonite, characterized by partial non-swelling and high swelling behavior, respectively [16].

Typically, determining the geotechnical characteristics of soft soils requires significant resources in terms of effort, time, labor, instruments, and finances. In order to get accurate data on compaction characteristics and soil unconfined compressive strength (*UCS*), a minimum of six tests and four experiments need to be conducted, respectively. In order to tackle this issue, predictive algorithms were developed to accurately evaluate compaction properties by employing *UCS* and other soil constituents. Data-driven based models have been utilized to extend a highly accurate prediction system for most soil properties, while the accuracy rate was low for big samples [19–24]. It utilized the Adaptive Neuro-Fuzzy Inference System (*ANFIS*) to compute the mechanical properties of processed sand mixed with fly-ash and cement [25]. Tree-based metaheuristics models were utilized to forecast the *UCS* of clayey soil that was modified with geopolymer [26]. The suggested model incorporates many factors that have a direct impact on the *UCS* of soil, including the plasticity index, plastic limit, additive quantities, and so on. In addition, Support Vector Regression (*SVR*) was used to calculate the mechanical properties of soil treated with a geopolymer-based admixtures. The results illustrated that the *SVR* was able to precisely estimate the shear strength of soil that had been enhanced with geopolymer. Güllü and Fedakar [27] have utilized several machine learning models to determine the *UCS* of soil that had been improved with fly-ash, steel fiber, and jute fiber. All of the algorithms examined yielded outcomes that exhibited a strong correlation with the actual *UCS* value of the soil. The forecasting of *UCS* of *MC* altered with recycled tiles (*RT*) was the main focus of the research. Four unique hybridized algorithms were created to achieve this goal by combining the Aquila optimizer (*AO*) algorithm with *ANFIS*, *SVR*, random forest (*RF*), and extreme gradient boosting (*XGB*). The results demonstrated that the *UCS* estimation process in all four models had dependable accuracy. Of all models, $AO - XGB$ approach has the lowest quantity of scatter index (*SI*) than others (the best efficiency), with $SI_{Train} = 0.0247$ and $SI_{Test} = 0.02483$. With 9.735 and 1.44 for the training and testing dataset, respectively, it was evident that the $AO - XGB$ model had the least amount of uncertainty, indicating its superior generalization ability over the other models [28].

The study employed both simple and hybrid *SVR* models to assess the *UCS* of bricks waste composites containing various amounts of peat [29]. Additionally, the *SVR* algorithms were used to estimate the *UCS* of improved soils [30], determine the strength of geopolymer-stabilized clayey soil [31], estimate the *UCS* of consolidated pond ashes [32], and predict the *UCS* of mixtures containing cockle shell, cement, and sand [33], among other applications. Furthermore, the use of RF was also implemented for the purpose of estimation, namely in the field of *UCS* prediction for soil-cement combination [34], *UCS* assessment for rocks [35], and calculation of undrained shear strength of soil using *RF* and *PSO* [36]. Nevertheless, doing research that incorporates the LSSVR framework and hybridizes it with several newly discovered optimization methods might be seen as an appealing innovation.

Regarding the mentioned gap on this topic, the present research focused on estimating the *UCS* of *MC* modified using *RT* with novel, accurate, and reliable frameworks. For this aim, the powerful and dependable algorithms named least square support vector regression (*LSSVR*) analysis was considered by introducing percent and size of recycled tiles (*PRT* and *SRT*), curing time (*CT*), highest dry density (*MDD*), optimum water content (*OMC*), *PH*, and index of plasticity (*PI*) of *MC* as input variables. A comprehensive dataset was attempted to be collected from previously published resources related to the target. It is worth mentioning that the effectiveness and accuracy of the *LSSVR* model are extremely concentrated on its optimal values hyperparameters. Where in the present study, three cutting-edge optimization algorithms linked with *LSSVR* in order to tune them, such as golden jackal algorithm (*GJA*), artificial rabbit optimizer (*ARA*), and dwarf mongoose algorithm (*DMA*). Finally, the sensitivity study was conducted so as to determine the effect of input parameters on the output.

10.2 Methodology

10.2.1 Dataset Pre-processing

This experiment employed the problematic *MC* modified with the recycled tile debris (*RT*). Both the modified and unmodified *MC* instances underwent Atterberg limits tests, which were conducted to assess changes in property behavior [37]. The compaction experiments were conducted on both the modified and unmodified *MC* to assess the impact of *RT* on compaction parameters, namely *MDD* and *OMC* [38]. Twelve samples were analyzed to assess the *UCS* of *MC* at four different curing times: 0 days, 7 days, 14 days, and 28 days. Under circumstances where the strain was controlled, the *UCS* was determined for a total of 156 samples, both modified and unmodified [1, 39, 40]. The study examined many parameters that might potentially impact the performance, including the dimensions and proportions of *RT* (*SRT* and *PRT*), the period of curing (*CT*), and the mechanical properties of *MC*. The objective was to develop reliable algorithms for forecasting the *UCS* of both stabilized and unstabilized *MC*. During the *UCS* calculation, the following characteristics were deemed effective in the group of mechanical characteristics: (*PI*), *MDD*, *OMC*, *PH*, liquid limit (*LL*), and plastic limit (*PL*) [1, 25, 40, 41]. For the reason of modeling, the collected data collection (156 samples) from the literature was randomly discreted into two stages: training data, which made up 75% of the total (117), and testing data, which made up 25% (39). Before this step, the quality of the dataset was analyzed by considering out-of-range data, discrete data, and their distribution. Table 10.1 and Fig. 10.1 indicate the statistical properties of the introduced attributes and the target, along with their circular distribution.

The Pearson correlation coefficient (*R*) is a statistical index that gauges the efficiency and direction of a linear relationship between two continuous variables [42]. The formula to calculate the *R* between two variables, *Z* and *P*, with *n* data points, is expressed as Eq. (10.1).

Table 10.1 Statistical properties of the introduced attributes and the target

Attributes	Subset	Index							
		Min.	*Max.*	*St. D.*	*Skew.*	*Range*	*Kurt.*	*Avg.*	*Med.*
1st input: *PRT* (%)	*Train*	0	40	12.238	−0.2875	40	−1.03	24.957	30
	Test	0	40	12.1355	0.3409	40	−0.8158	17.436	20
2nd input: *SRT* (mm)	*Train*	0	0.3	0.10174	0.3038	0.3	−1.34	0.16095	0.15
	Test	0	0.3	0.1119	0.2885	0.3	−1.4175	0.14854	0.15
3rd input: *CT* (days)	*Train*	0	28	10.0602	0.5176	28	−0.98	11.7265	7
	Test	0	28	11.0381	0.2073	28	−1.5308	13.8205	14
4th input: *MDD* (kg m^{-3})	*Train*	1.59	1.77	0.04212	−0.9935	0.18	1.129	1.70966	1.72
	Test	1.59	1.77	0.05111	−0.417	0.18	−0.4162	1.6834	1.69
5th input: *OMC* (%)	*Train*	15	22	1.5228	0.696	7	1.402	17.706	18
	Test	15	22	1.856	0.6103	7	0.07124	18.421	18
6th input: *PH*	*Train*	2.61	2.89	0.063	1.2613	0.28	2.173	2.6987	2.69
	Test	2.61	2.89	0.075	0.9511	0.28	0.2433	2.7377	2.73
7th input: *PI* (%)	*Train*	10.57	19	2.111	0.9396	8.43	0.558	13.7465	13.15
	Test	10.57	19	2.8126	0.336	8.43	−1.468	14.9267	13.94
Target: *UCS* (kPa)	*Train*	50	232	42.91	−0.3788	182	−0.01	152.068	158
	Test	50	237	56.357	−0.323	187	−1.0423	143.41	154

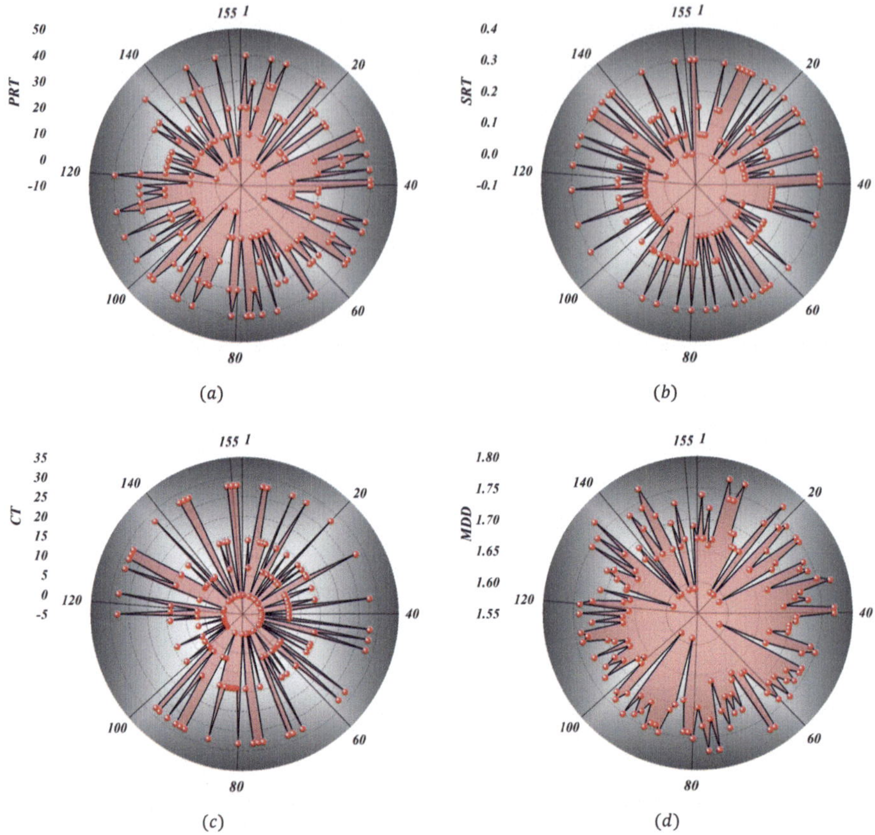

Fig. 10.1 The circular visualization of the attributes; a) PRT, b) SRT, c) CT, d) MDD, e) OMC, f) PH, g) PI, h) UCS

$$R = \Sigma\left(Z_i - \bar{Z}\right)\left(P_i - \bar{P}\right) / \sqrt{\Sigma\left(Z_i - \bar{Z}\right)^2 \times \Sigma\left(P_i - \bar{P}\right)^2} \quad (10.1)$$

where Z_i and P_i illustrate single data points, and $\bar{Z}$ and $\bar{P}$ are the means of the Z and P variables, respectively. The R spans from −1 to +1, where R at 1 denotes a complete positive linear correlation (stronger linear relationship), R at −1 presents a complete negative linear relationship, and R at 0 signifies no linear correlation. A positive R implies a positive relationship, indicating that as one variable enhances, the other tends to enhance. A negative R signifies a negative relationship, where an enhance in one variable attributes to a decrease in another one. The assumptions of this analysis are that the relationship among parameters should be linear and that parameters should be approximately normally divided. The results are presented in Fig. 10.2. As it is obvious, the parameter *CT* demonstrated zero R with other parameters, with the exception of *UCS* at 0.597. All of the attributes contain the moderate R value with the target *UCS*. The highest positive R assigned between *PRT* and

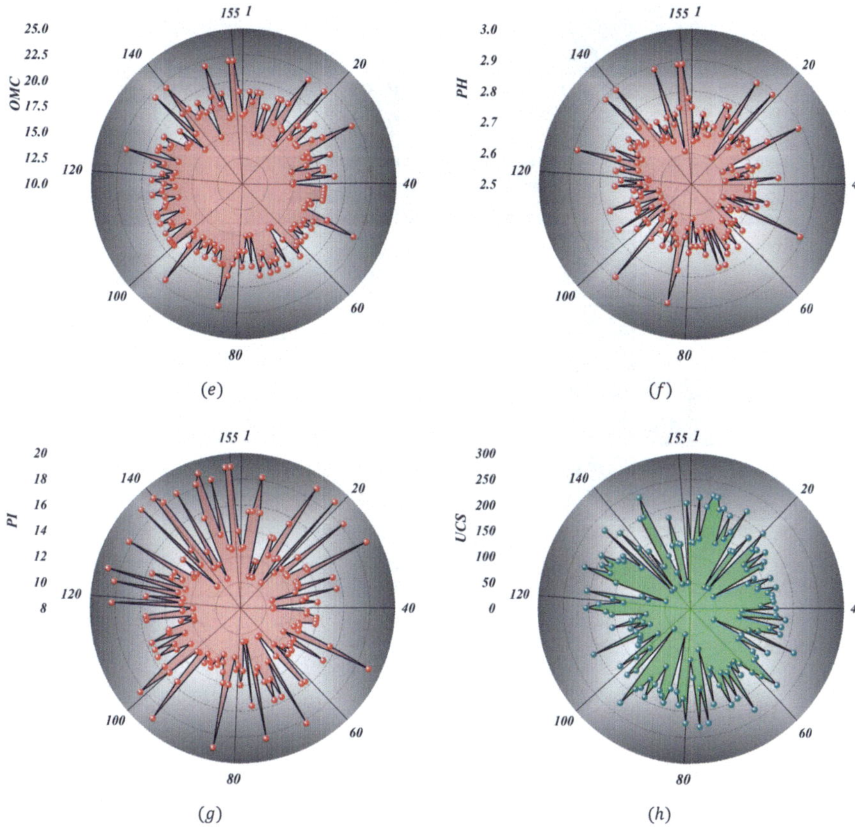

Fig. 10.1 (continued)

MDD at 0.857, whereas the negative *R* was related between *MDD* and *OMC* by −0.871. Therefore, the developed frameworks must be powerful to obtain accurate results overcoming these high values of correlations.

Fast Sensitivity Analysis (*FAST*) is a method used to determine how the variability in the output of a model can be attributed to different sources of variability in the input parameters. It is particularly useful in finding the impact of each input variable on the model's output, especially when dealing with complex models with multiple inputs. This involves assessing how sensitive a model's output is to changes in its input parameters. This is a specific technique within sensitivity analysis that uses Fourier series to represent the input-output relationship of the model.

FAST can handle approaches with a most of input parameters more efficiently compared to other methods like Monte Carlo simulation. Also, it provides a clear understanding of the relative significance of each input variable. Moreover, the method is less prone to sampling errors and can provide reliable sensitivity measures.

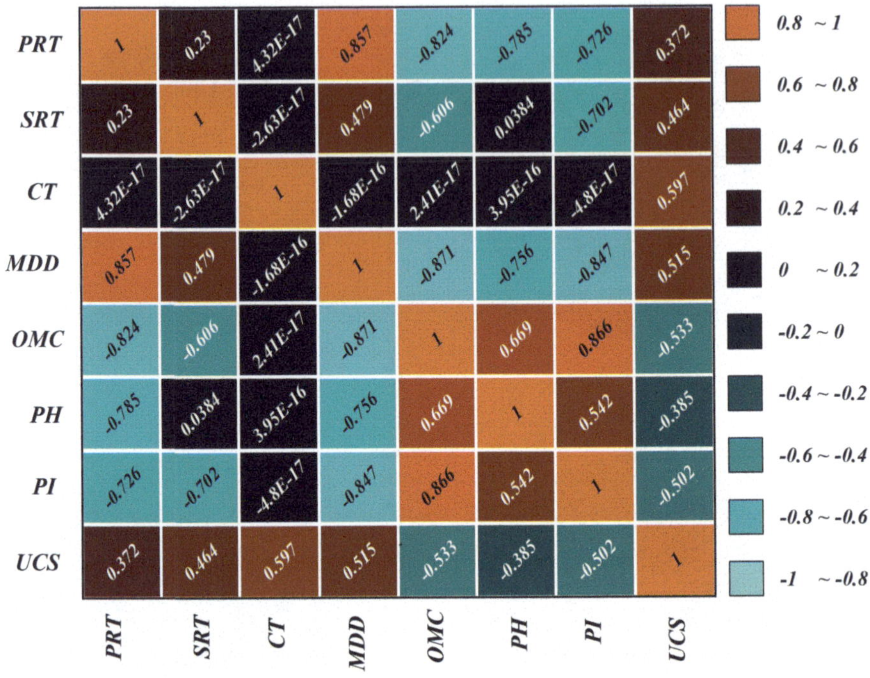

Fig. 10.2 The coefficient of Pearson correlation

In the context of sensitivity analysis, specifically within the framework of *FAST* and other variance-based methods, S_1 and S_T are sensitivity indices that provide insights into the effect of input variables on the output of a model. S_1 shows first-order sensitivity indice, estimates the direct influence of an input variable on the output variance, assuming all other parameters are fixed. It is calculated by integrating the variance of the output with respect to the input parameter of interest while keeping other parameters constant. On the other hand, S_T, or the entire effect indice, estimates the entire contribution of an input parameter to the output variance, containing both its direct effects and all higher order relationships with other input variables. A high S_1 value represents that the input parameter X_i has a high direct effect on the output. A high S_T value means that X_i has a high effect on the output, including its interactions with other parameters. The results of this analysis are provided in Fig. 10.3. As it can be seen, considering the values of S_1, the *PH* parameter depicted the highest value at 0.6275, followed by *MDD* at 0.5217. Conversely, the *OMC* parameter received the lowest value at 0.1268 shows its lower impact on the target. On the other hand, regarding S_T, the importance of *MDD* is incredibly high (roughly one) by 0.9999. Also, the second and third highest important attributes are *PI* and *PH* at 0.9874, and 0.9807, respectively.

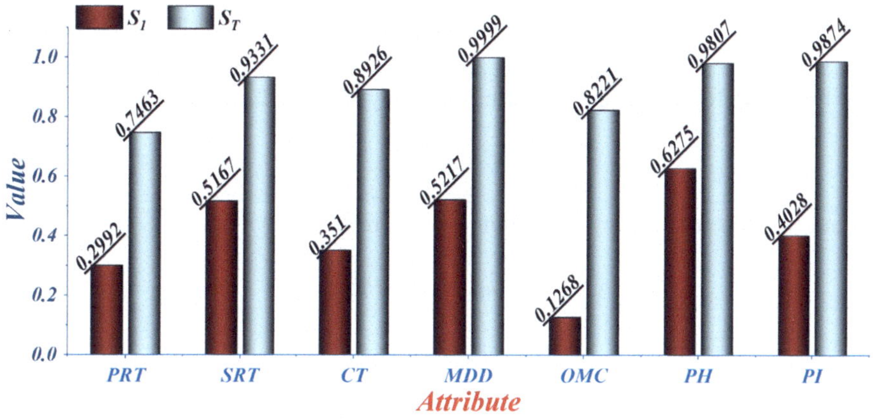

Fig. 10.3 The results of feature importance

10.2.2 *Golden Jackal Algorithm* (GJA)

The GJA approach, an optimization technique inspired by nature, was developed by Chopra and Ansari [43]. GJA offers an extra optimization technique for addressing practical engineering problems, similar to many other metaheuristic approaches. GJA is based on the cooperative hunting behavior of golden jackals. Jackals hunt in groups or in couples, with the male and female traveling together. When golden jackals hunt in pairs, they locate their prey, approach it, agitate, and encircle it until it stops moving, and then leap on it. Golden jackals have two distinct stages in their social hunting process: (a) exploration, which entails following, recognizing, and locating the hunt, and (b) exploitation, which entails enclosing and hitting the hunt. The mathematical models for the two procedures are explained in the following section.

10.2.2.1 *GJA* Mathematical Model

The mathematical models created by the different groups are shown in this section. Finding, identifying, tracking, and attacking the target are explained.

Phase of discovery (finding and pursuing the prey): While hunting, the male jackal is accompanied by the female jackal. The updated locations were obtained by applying Eqs. (10.2) and (10.3) to the male and female jackals. The locations of the male and female jackals are represented by $P_M(t)$ and $P_{FM}(t)$, respectively, where t indicates the number of iterations in the current iteration and $Prey(t)$ indicates the prey's location vector. According to the hunt, male and female jackals can currently be found at $P_1(t)$ and $P_2(t)$.

$$P_1(t) = P_{1M}(t) - E \cdot \left| P_M(t) - rl \cdot Prey(t) \right| \tag{10.2}$$

$$P_2(t) = P_{FM}(t) - E \cdot \left| P_{FM}(t) - rl \cdot Prey(t) \right| \tag{10.3}$$

The following formula is used to determine Prey Evading Energy, or E:

$$E = E_1 * E_0 \tag{10.4}$$

E_1 and E_0 stand for the prey's dropping energy, respectively.

$$E_1 = m_1 * \left(1 - (t / T)\right) \tag{10.5}$$

$$E_0 = 2 * r - 1 \tag{10.6}$$

In this case, m_1 is the constant equal to 1.5, t is the current iteration, and T is the maximum number of iterations. E_1 falls linearly from 1.5 to 0 after each cycle, while r is randomly selected between 0 and 1. In Eqs. (10.2) and (10.3), where the vector of accidental integers is obtained from the Lévy distribution, the Lévy movement is denoted by (rl). The combination of rl and prey simulates the movement of the prey in a Lévy fashion.

$$rl = 0.05 * LF(p) \tag{10.7}$$

The following formula can be used to calculate the Lévy flight function.

$$LF(p) = 0.01 \times \frac{(\mu + \delta)}{\left(\left| v^{\left(\frac{1}{\beta}\right)} \right| \right)}; \left(\frac{\Gamma(1+\beta) \times \sin(\pi\beta / 2)}{\Gamma\left(\frac{1+\beta}{2}\right) \times \beta \times 2^{\frac{\beta-1}{2}}} \right)^{1/\beta} \tag{10.8}$$

where u and v stand for accidental values in (0, 1) and the β parameter is set to 1.5. The final revision of the jackal locations is made by applying the average of Eqs. (10.2) and (10.3).

$$P(t+1) = \frac{P_1(t) + P_2(t)}{2} \tag{10.9}$$

Exploitation phase (pouncing and enclosing the prey): The hunt's capacity to escape is reduced when it is harassed by jackals during the exploitation phase (pouncing and encircling the prey), and the jackal pair encircles the hunt that was initially spotted during the exploration stage. They capture their victim, enclose it, and then eat it. The way that male and female jackals hunt together is explained by the following mathematical model:

$$P_1(t) = P_M(t) - E \cdot \left| rl \cdot P_M(t) - Prey(t) \right| \tag{10.10}$$

$$P_2(t) = P_{FM}(t) - E \cdot \left| rl \cdot P_{FM}(t) - Prey(t) \right| \tag{10.11}$$

where $P_1(t)$ and $P_2(t)$ indicate the updated locations of the male and female jackal pair during the process of surrounding and jumping on the hunt. The *GJA*'s primary steps are shown in Algorithm 10.1.

Algorithm 10.1: The GJA Pseudo-Code

✓	***Inputs:*** *The population size N and highest iterations T*
✓	***Target:*** *The position of prey and its best value*
✓	*Initialize the random prey population P_i (i = 1, 2, …, N)*
✓	***while*** *($t < T$)*
✓	*Compute the fitness values of preys*
✓	*P_1 =good prey (Male Jacal location)*
✓	*P_2 =Second good prey (Female Jacal location)*
✓	**for** each prey
✓	***Readjust*** *the evading energy E by Eqs. (10.4)–(10.6)*
✓	***Readjust*** *rl by Eqs. (10.7)–(10.8)*
✓	***if*** $\|E\| \geq 1$ # *Exploration stage*
✓	***Readjust*** *the prey location Eqs. (10.2)–(10.3) and (10.9)*
✓	***if*** $\|E\| < 1$ *#Exploitation stage*
✓	***Readjust*** *the prey location Eqs. (10.10)–(10.11) and (10.9)*
✓	**end for**
✓	*T=t+1*
✓	***end while***
✓	***Return*** Y_1

10.2.3 Dwarf Mongoose Algorithm (DMA)

DMOA is an *MH* technique that was developed to replicate certain aspects of the dwarf mongoose's behavior, including its restrictions on prey size, social structure, semi-nomadic lifestyle, and more [44]. The first step in the *DMA* is to initialize a collection of arbitrary candidate mongoose populations among the issue's minima (*LB*) and maxima (*UB*). The subsequent stages reflect the optimization procedure:

10.2.3.1 Alpha Group

Every group's fitness likelihood may be calculated upon initializing the population by:

$$\alpha = \frac{fit_i}{\sum_{i=1}^{n} fit_i} \tag{10.12}$$

A route *peep* indicated by $\alpha - bs$ number of vocalizations from alpha females keeps the family inside it for *bs* number of babysitters. The starting value of the sleeping mound is set to α while each mongoose is asleep. *DMA* utilizes the following methods to create a potential food location:

$$X_{i+1} = x_i + phi + peep \tag{10.13}$$

In this section, *phi* represents arbitrarily selected from a uniform distribution that spans from 0 to 1. Following each iteration, the slumber mound *Sm* may be expressed as follows:

$$sm_i = \frac{fit_{i+1} - fit_i}{\max\left[fit_{i+1}, fit_i\right]} \tag{10.14}$$

The representation of the *SM*'s average (*Avg*) value is as follows:

$$Avgsm = \frac{\sum_{i=1}^{n} sm_i}{n} \tag{10.15}$$

The *DMA* is subsequently transferred to the scouting step. Following the fulfillment of the babysitter exchange condition, the slumber mound, or the subsequent food resource is evaluated during this stage.

10.2.3.2 Scout Group

This stage involves observing the subsequent sleeping mound, which ensures exploration is assured due to the mongooses' reluctance to revisit the previous sleeping mound. This movement is determined by the Mongooses' overall performance. Motivating this is the possibility that substantial foraging will unveil a novel *sm*. The scout mongoose may be shown as follows:

$$X_{i+1} = \begin{cases} X_i - CF * phi * rand\left[X_i - \vec{M}\right] & if\ Avgsm_{i=1} > Avgsm_i \\ X_i + CF * phi * rand\left[X_i - \vec{M}\right] & otherwise \end{cases} \tag{10.16}$$

Here, $CF = \left(1 - \frac{iter}{\max_{iter}}\right)^{2\frac{iter}{\max_{iter}}}$ and $\vec{M} = \sum_{i=1}^{n} \frac{X_i * sm_i}{X_i}$. The mongoose moves to the novel *sm* based on a vector called X_i, and its collective-volitive movement is controlled by a variable called *CF*. The number of iterations reduces *CF* in a linear manner. A uniform distribution over [0, 1] is used to arbitrarily select the *rand*, and $\vec{M}$ represents a vector that presents the mongoose's progress to the newer *sm*.

10.2.3.3 The Babysitters

The babysitters, who are supportive members of the crew, look after the young mongooses. A systematic rotation of babysitters enables the alpha female (mother) to freely search for food on a daily basis while the rest of the crew takes care of the offspring. During both noon and evening, the alpha female comes back to nurse the offspring. The population size determines the number of babysitters. The proportion of babysitter representation emulates this group by reducing the population, hence impacting the *DMA*. The exchange variable of the babysitter resets the food supply and scouting information previously held by the replacement members of the family. In the subsequent iteration, the mean weight of the alpha group is decreased by assigning a fitness weight of zero to the babysitters. It highlights exploitation and impedes collective mobility. The steps of the *DMA* are provided in Algorithm 10.2.

Algorithm 10.2: The DMA Pseudo-Code

✓	*Initialize the parameter:* [*peep*], *the population: n, the number of babysitters: bs*
✓	*Set* $n = n - bs$
✓	*Set babysitters exchange parameter L*
✓	***for*** $iter = 1 : max\_iter$
✓	*Compute the fitness of the mongoose*
✓	*Set time counter C*
✓	*determine the alpha by Eq. (10.12)*
✓	*Create a candidate food location by Eq. (10.13)*
✓	*assess new fitness of* X_{i+1}
✓	*assess sleeping mound by Eq. (10.14)*
✓	*Calculate the mean value of the sleeping mound found by Eq. (10.15)*
✓	*Calculate the movement vector by:* $\vec{M} = \sum_{i=1}^{n} \frac{X_i \times sm_i}{X_i}$
✓	*Exchange babysitters if* $C \geq L$*, and set*
✓	*Initialize bs location (Eq. 10.1) and compute fitness* $fit_i \leq \alpha$
✓	*Simulate the scout mongoose next position using Eq. (10.16)*
✓	*Readjust best answer so far*
✓	***end for***
✓	***return*** *best answer*
✓	***end***

10.2.4 Artificial Rabbit Algorithm (**ARA***)*

The *ARA* method was inspired by the natural survival techniques employed by rabbits [45]. The detour foraging technique is the process through which rabbits search for food far away from their homes. They dig tunnels surrounding their nests and use the random hiding strategy—which involves hiding in one of them at random—to avoid predators and attackers. Depending on their energy level, they will decide between random concealment and detour foraging. When they are highly or sufficiently energetic, they will forage for food at a distance from their nests (a process known as detour foraging), and when they are short on energy, they will accidentally conceal in the nearby burrows.

10.2.4.1 Energy Shrink (Switch Between Exploration and Exploitation)

Rabbits may decide to forage on detours or hide at random. In order to imitate which one the rabbit would choose, an energy factor *A*(*t*) is determined using Eq. (10.17) based on the quantity of energy the rabbit possesses. A rabbit will forage on diversion when $A(t) > 1$ occurs and will randomly hide when $A(t) \leq 1$ occurs.

$$A(t) = 4\left(1 - \frac{t}{T}\right) in \frac{1}{r} \tag{10.17}$$

Here, *r* represents a randomly selected integer in the interval (0, 1).

10.2.4.2 Detour Foraging (Exploration)

In order to keep attackers away from their nests, rabbits forage for food elsewhere. Based on Eq. (10.18), rabbits haphazardly look for food based on one another's positions.

$$\begin{aligned} \vec{p}_i(t+1) &= \vec{x}_j(t) + R \times \left(\vec{x}_i(t) - \vec{x}_j(t)\right) \\ &+ round\left(0.5 \times \left(0.05 + r_2\right)\right) \times n_1, \quad i, j = 1, \ldots, n \text{ and } j \neq i \end{aligned} \tag{10.18}$$

$$R = L \times c \tag{10.19}$$

$$L = \left(e - e^{\left(\frac{t-1}{T}\right)^2} \times \sin\left(2\pi r_2\right)\right) \tag{10.20}$$

$$c(k) = \begin{cases} 1 \text{ if } k = g(l) \\ 0 \text{ else} \end{cases} \quad k = 1, \ldots, d \text{ and } l = 1, \ldots, \left[r_3 \cdot d\right] \tag{10.21}$$

$$g = rand\ perm(d) \tag{10.22}$$

$$n_1 \sim N(0,1) \tag{10.23}$$

Here, n represents the size of the rabbit population, d denotes the number of parameters in the issue that require to be optimized, T shows the highest number of iterations, L stands for the rabbits' motion speed, r_1, r_2, and r_3 denote three arbitrary numbers among (0, 1), and n_1 is subject to the standard normal distribution. Where $\vec{p}_i(t+1)$ represents the candidate's location of the ith rabbit at the time $t + 1$, while $\vec{x}_i(t)$ defines the ith rabbit's location at the time t. R provides the running operator that mimics the running features of rabbits, and c represents a mapping vector.

10.2.4.3 Random Hiding (Exploitation)

Every rabbit has a number of tunnels surrounding its nest from which it may arbitrarily choose one to conceal in order to elude attackers. Equation (10.24) generates these tunnels for every rabbit.

$$\vec{b}_{i,j}(t) = \vec{x}_i(t) + H \times g \times \vec{x}_i(t), \quad i = 1,\ldots,n \text{ and } j = 1,\ldots,d \tag{10.24}$$

$$H = \frac{T - t + 1}{T} \times r_4 \tag{10.25}$$

$$g(k) = \begin{cases} 1 \text{ if } k = j \\ 0 \text{ else} \end{cases} k = 1,\ldots,d \tag{10.26}$$

$$\vec{p}_i(t+1) = \vec{x}_i(t) + R \times \left(r_4 \times \vec{b}_{i,r}(t) - \vec{x}_i(t)\right), \quad i = 1,\ldots,n \tag{10.27}$$

$$g_r(k) = \begin{cases} 1 \text{ if } k = [r_5 \times d] \\ 0 \text{else} \end{cases} k = 1,\ldots,d \tag{10.28}$$

$$\vec{b}_{i,r}(t) = \vec{x}_i(t) + H \times g_r \times \vec{x}_i(t), \quad i = 1,\ldots,n \tag{10.29}$$

$$\vec{x}_i(t+1) = \begin{cases} \vec{x}_i(t) f(\vec{x}_i(t)) \le f(\vec{p}_i(t+1)) \\ \vec{p}_i(t+1) f(\vec{x}_i(t)) > f(\vec{p}_i(t+1)) \end{cases} \tag{10.30}$$

where r_4 and r_5 are random values among 0 and 1, H represents the concealing variable, $\vec{b}_{i,j}$ indicates the jth burrow for the ith rabbit, and $\vec{b}_{i,r}$ denotes an arbitrarily picked burrow for concealing for the ith rabbit, as indicated in Eq. (10.29). Based on Eq. (10.27) the ith rabbit will try to adjust its posture in accordance with the burrow that is arbitrarily selected. Equation (10.30) finally demonstrates that the rabbit will abandon its present location and stay at the candidate's location after either arbitrary concealment or detour foraging, provided that the fitness of the candidate's location of *i*th rabbit is higher than that of the preceding one. Algorithm 10.3 presents where the *ARA* starts and where finishes.

Algorithm 10.3: The ARA Pseudo-Code

✓	*Randomly assign a group of rabbits X_i and assess their fitness (Fit), with X_{best} being the currently discovered optimal solution.*
✓	***while*** *the stopping circumstances is not gotten,*
✓	***for*** *each individual X_i, perform the below terms:*
✓	*Calculate the energy factor A by Eq. (10.17)*
✓	***if*** $A > 1$
✓	*Choose an individual rabbit randomly*
✓	*Identify R by Eqs. (10.19)–(10.23)*
✓	*Perform detour foraging by Eq. (10.18)*
✓	*Calculate the fitness Fit_i*
✓	*Set the current individual's location by Eq. (10.30)*
✓	***else***
✓	*Produce d burrows and randomly choose one for hiding by Eq. (10.29)*
✓	*Perform random hiding by Eq. (10.27)*
✓	*Calculate the fitness Fit_i*
✓	*Readjust the individual's location by Eq. (10.30)*
✓	***end if***
✓	*Readjust the best location found thus far, X_{best}*
✓	***end for***
✓	***end while***
✓	***return*** X_{best}

10.2.5 Least Square Support Vector Regression (LSSVR)

In accordance with the statistical theory of learning, support vector machine (*SVM*) is a popular discriminant approach [46, 47]. It is according to the idea of structural risk minimization and may be applied to a wide range of data types. Furthermore, a linearly constrained quadratic programming issue is identical to *SVM* training. Consequently, in comparison with neural network training, *SVM* training allows for the search for the specific optimum answer and makes the training process more difficult to reach the local minimum. Nonetheless, the quantity of samples and the number of restrictions in this issue are the same. As a result, *SVM* will need a lot of calculation and effort for training sets, including lots of data. *LSSVR* is designed to solve the laborious restricted quadratic programming issue. It is according to the ordinary *SVR*. *LSSVR* uses the equality constraint rather than the *SVR*'s inequality requirement and inserts the square function of the error into the objective function. As a result, the training pace is comparatively quick, and the particular method is as follows. The definition of the training sample set includes *N* pairs of data, as shown below:

$$S = \left\{ s_i \middle| s_i = (x_i, y_i), x_i \in R^n, y_i \in R \right\}_{i=1}^{N} \tag{10.31}$$

where y_i represents the actual output value and x_i denotes the n-dimension input vector. With the regression issue in mind, the *LSSVR* approximation function is provided as follows:

$$y = W^T \cdot \varphi(x) + b \tag{10.32}$$

Here, w stands for the weight vector in reproducing kernel Hilbert space, b represents the bias term, and $\varphi(x)$ denotes a nonlinear function that transfers the input vector from the source field to *RKHS*. Equation (10.33) illustrates how the issue of Eq. (10.32) is converted into a constraint optimization issue in accordance with the notion of reducing structural risk.

$$\begin{cases} \min\limits_{W,b,e} f(W,e) = \frac{1}{2} W^T W + \frac{1}{2} C \sum\limits_{i=1}^{N} e_i^2 \\ \text{subject to}: y_i = W^T \cdot \varphi(x_i) + b + e_i \end{cases} \tag{10.33}$$

Here, the regularization factor that strikes an equilibrium among the model's intricacy and precision is denoted by C, the object function, and the outcome error among the model and actual value, represented by f and e, respectively. The object function is convex because its Hessian matrix is positive. The best answer for the convex square programming issue may be found using Lagrange multipliers. The following is an example of the Lagrange function:

$$\mathcal{L}(w,b,e_i,\alpha_i) = \frac{1}{2} W^T W + \frac{1}{2} C \sum_{i=1}^{N} e_i^2 - \sum_{i=1}^{N} \alpha_i \left[W^T \varphi(x_i) + b + e_i - y_i \right] \tag{10.34}$$

The Lagrange multipliers are represented by α_i. Suykens et al. [51] state that the Karush-Kuhn-Tucker (*KKT*) criteria allow for the transformation of the optimization issue into Eq. (10.35), that is demonstrated in the following manner:

$$\begin{cases} \frac{\partial L}{w} = 0 \rightarrow w = \sum\limits_{i+1}^{N} \alpha_i \cdot \varphi(x_i) \\ \frac{\partial L}{\partial b} = 0 \rightarrow \sum\limits_{i+1}^{N} \alpha_i = 0 \\ \frac{\partial L}{\partial e_i} = 0 \rightarrow \gamma e_i = \alpha_i \quad i = 1, \ldots, N \\ \frac{\partial L}{\partial \alpha_i} = 0 \rightarrow y_i = W^T \varphi(x_i) + b + e_i \quad i = 1, \ldots, N \end{cases} \tag{10.35}$$

To make things simpler, w and α_i might take the place of φ and e_i. After that, we shall express Eq. (10.35) in the following manner:

$$\begin{bmatrix} 0 & 1 & \dots 1 \\ 1 & K+\dfrac{1}{C} & \dots & K \\ \vdots & \vdots & \ddots & \vdots \\ 1 & K & \dots & K+\dfrac{1}{C} \end{bmatrix} \begin{bmatrix} b \\ \alpha_1 \\ \vdots \\ \alpha_N \end{bmatrix} = \begin{bmatrix} 0 \\ y_1 \\ \vdots \\ y_N \end{bmatrix} \tag{10.36}$$

When the kernel function, denoted by $K = \varphi^T(x)\varphi(x)$, is guaranteed to meet the Mercer condition. Equation (10.36), when solved, yields w and α_i. Thus, the following is how the ultimate output result was stated:

$$f(x) = \sum_{i=1}^{N} \alpha_i K(x_i, x) + b \tag{10.37}$$

According to the derivation, the penalty factor (C) and kernel function width (*Epsilon*) directly affect *LSSVR* regression results. In *LSSVR*, the C controls the trade-off between fitting the training data well and keeping the model simple. A high value of C puts more emphasis on fitting the training data precisely, aiming for a smaller training error. High C values may lead to overfitting, where the model captures noise and fluctuations in the training data. The resulting model may not generalize well to unseen data. Also, a higher C allows the model to become more complex, potentially capturing the noise in the training data. Conversely, a low value of C allows for a more flexible model with a larger margin for error on the training data. Very low C values may lead to underfitting, where the model is too simple to capture the underlying patterns in the data. Moreover, lower C values promote a smoother model, which may generalize better to unseen data.

Turning to *Epsilon*, this parameter is often associated with radial basis function (*RBF*) kernels, where the width is controlled by the hyperparameter. A higher *Epsilon* makes the influence of each data point more localized. The contribution of each support vector diminishes more quickly with distance from the vector. A wider kernel leads to smoother predictions and potentially a smoother decision surface. Extremely high values of *Epsilon* might lead to underfitting, where the model may become too simple and fail to capture intricate patterns in the data. Conversely, a lower *Epsilon* increases the influence of each data point over a larger region. Smaller kernel widths can help the model capture local patterns and details in the data, making it potentially more sensitive to individual data points. Extremely low values of *Epsilon* might lead to overfitting, where the model may fit noise and fluctuations in the training data. The described hyperparameters could be determined by trial-and-error method or using optimization algorithms. Correspondingly, this research determined the optimal values of these hyperparameters by employing *GJA*, *ARA*, and *DMA*.

10.2.6 Evaluation Indices

Evaluation indices serve the purpose of assessing the performance and effectiveness of models. The primary aim of these indices is to provide quantitative measures that understand how well a model is performing in relation to its intended tasks. Performance assessment, comparison of models, generalization assessment, and robustness testing are some of the key aims of using evaluation indices. The considered metrics in the present study are the coefficient of determination (R^2), root mean square error (*RMSE*), normalized *RMSE* (*NRMSE*), mean absolute error (*MAE*), Theil inequality coefficient (*TIC*), indices of agreement (*IA*), normalized mean absolute error (*NMSE*), uncertainty with 95% confidence level (*Un*(95)), scatter index (*SI*), and a comprehensive index called *OFU*. These metrics have been widely used in similar studies [48–50].

$$R^2 = \left(\frac{\sum_{g=1}^{G} \left(n_g - \overline{n}\right)\left(z_g - \overline{y}\right)}{\sqrt{\left[\sum_{g=1}^{G} \left(n_g - n\right)^2\right]\left[\sum_{g=1}^{G} \left(y_g - \overline{y}\right)^2\right]}} \right)^2 \tag{10.38}$$

$$RMSE = \sqrt{\frac{1}{G} \sum_{g=1}^{G} \left(y_g - n_g\right)^2} \tag{10.39}$$

$$NRMSE = RMSE \,/\, \overline{y} \tag{10.40}$$

$$MAE = \frac{1}{G} \sum_{g=1}^{G} \left| y_g - n_g \right| \tag{10.41}$$

$$TIC = \frac{\sqrt{\frac{1}{G} \sum_{g=1}^{G} \left(y_g - n_g\right)^2}}{\left(\sqrt{\frac{1}{G} \sum_{g=1}^{G} y_g^{\;2}} + \sqrt{\frac{1}{G} \sum_{g=1}^{G} n_g^{\;2}} \right)} \tag{10.42}$$

$$IA = 1 - \frac{\sum_{g=1}^{G} \left(n_g - y_g\right)^2}{\sum_{g=1}^{G} \left(\left| n_g - \overline{n} \right| + \left| y_g - \overline{n} \right| \right)^2} \tag{10.43}$$

$$NMSE = MSE \,/\, Var\left(n\right) \tag{10.44}$$

$$Un(95) = 1.96\sqrt{\left(SD^2 + RMSE^2\right)} \tag{10.45}$$

$$SI = \frac{\sqrt{\left(\frac{1}{G}\right)\sum_{g=1}^{G}\left(\left(n_g - \bar{n}\right) - \left(y_g - \bar{y}\right)\right)^2}}{\left(\frac{1}{G}\right)\sum_{g=1}^{G} y_g} \tag{10.46}$$

$$OFU = \left(\frac{g}{G} \times \frac{RMSE + MAE}{R^2 + 1}\right)^{Train} + \left(\frac{g}{G} \times \frac{RMSE + MAE}{R^2 + 1}\right)^{Test} \tag{10.47}$$

where n_g is the recorded UCS_i, $\bar{n}$ is the average of UCS_i, y_g is the simulated UCS_i, and $\bar{y}$ is the average of the UCS_i. g_{train} and g_{test} show the number of rows in the train and test stages of the dataset with the total number of *G*. Here, *SD* shows the standard deviation.

10.3 Findings and Justifications

10.3.1 The Development Outlines

Developing an *LSSVR* model involves several key steps, including data pre-processing, model training, hyperparameter identification through optimization and evaluation. The outlined steps of the model's development are as follows. The data pre-processing action was performed, including data cleaning, handling missing values, and removing outliers if necessary. After that, the dataset was standardized and divided into training and testing sets. Following this, the initial *LSSVR* model was started with initialized values for hyperparameters (i.e., *C* and *epsilon*) with certain intervals. Next, cutting-edge optimization algorithms named *GJA*, *DMA*, and *ARA* were utilized to find the optimal values for hyperparameters. The *LSSVR* analysis was trained using a training set and assessed for testing sets applying the identified optimal hyperparameters. Next, appropriate regression metrics are considered and calculated as evaluation metrics to assess model performance. Finally, if necessary, additional hyperparameter tuning is performed based on evaluation results. The principal values of optimization algorithms, along with optimal values of the hyperparameters related to *LSSVR* models, are supplied in Table 10.2.

Table 10.2 The initialization of optimizers and tunned values related to each model

Optimization	Attribute	Value	*LSSVR* models	Optimal attribute	Value
GJA	Runs	20	*LSGJA*	*C* [1–200]	86
	Iterations	100		*Epsilon* [0.5–10]	5.08
	Populations	30			
	C_1	[1–2]			
ARA	Runs	20	*LSARA*	*C* [1–200]	101
	Iterations	100		*Epsilon* [0.5–10]	3.89
	Populations	50			
	–	–			
DMA	Runs	20	*LSDMA*	*C* [1–200]	63
	Iterations	100		*Epsilon* [0.5–10]	7.1
	Populations	35			
	Babysitters	3			
	peep	2			

10.4 Results

The subsequent results provide the findings of many hybridization estimating techniques used to analyze the *UCS* of *MC* modified with *RT*. For this aim, the powerful and dependable algorithms named least square support vector regression (*LSSVR*) analysis was considered. It is worth mentioning that the effectiveness and accuracy of the *LSSVR* model are extremely concentrated on its optimal values hyperparameters. Where in the present study, three cutting-edge optimization algorithms linked with *LSSVR* in order to tune them, such as golden jackal algorithm (*GJA*), artificial rabbit optimizer (*ARA*), and dwarf mongoose algorithm (*DMA*). Table 10.3 and Figs. 10.4 and 10.5 show the results of comparing the efficacy of the created algorithms to choose the best one. Ten measures for efficiency were used to assess the accuracy of the algorithms. Moreover, in order to prove the dependability of the frameworks, the findings of the present article were compared with developed models in the literature related to this topic by linking the *ANN* system with imperialist competitive algorithm (*ICA*) (*ICA* – *ANN*) [52].

The results indicate that the *LSGJA*, *LSARA*, and *LSDMA* methods have significant potential to accurately forecast the *UCS*. During the learning and examination stages, the R^2 and *IA* values for the *LSGJA* approach were found to be, respectively, 0.992 and 0.9962 and 0.998 and 0.999 (the greater, the better). With regard to *LSDMA*, these values were lower than those of *LSGJA*, with R^2 values of 0.9915 and 0.9936 and *IA* of 0.9978 and 0.9984, respectively. Despite having a strong prediction ability, the *LSSVR* algorithm optimized with *ARA* placed in third place by obtaining lower values. Other error-based measurements are also very helpful in determining if the method is valid. *RMSE*, *NRMSE*, *MAE*, *TIC*, *NMSE*, and *SI* are some of these measures where calculated in the present study. Though it is more appropriate than *LSARA*, it is evident from the values of these measures that *LSGJA* performs more effectively in comparison with *LSDMA*. For instance, with regard to

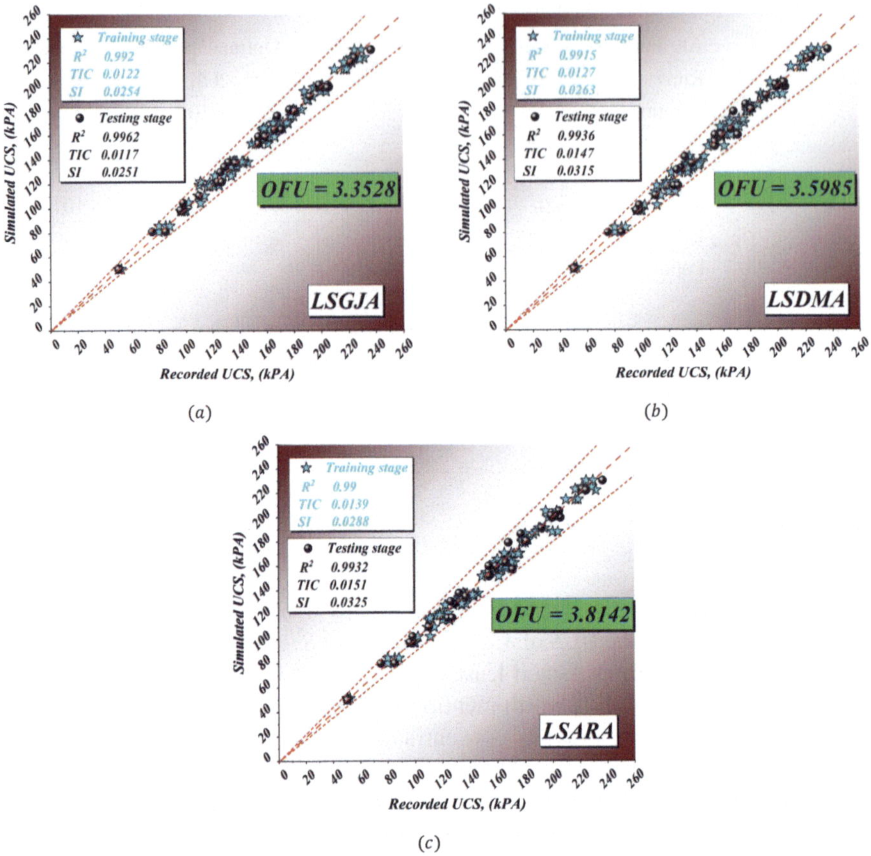

Fig. 10.4 The correlation plot of simulated and recorded; a) LSGJA, b) LSDMA, c) LSARA

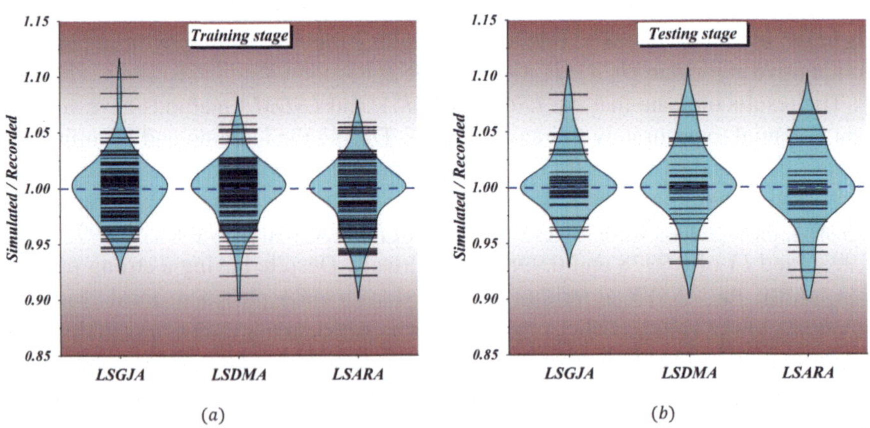

Fig. 10.5 The Simulated/Recorded Ration (*SRR*) distribution; a) training and b) testing

the values of the *TIC*, the slightest values owned by *LSGJA* at 0.0122 and 0.0117 in the learning and examining portions, respectively, slighter compared to *LSDMA* at 0.0127 and 0.0147, followed by 0.0139 and 0.0151 related to *LSARA*.

A range of numbers utilized to express the degree of uncertainty surrounding a prediction is called an uncertainty interval, also known as a confidence interval. It offers a gauge for the estimate's accuracy or dependability. Confidence intervals are often linked to a certain confidence level, such as 95% or 99%. The likelihood that the genuine parameter—such as a mean or proportion—falls within the estimated uncertainty range is represented by the confidence level. 95% was chosen for the confidence interval in the current investigation (the lower the numbers, the better the dependability). The *LSDMA* values for this index are in the center for the learning and examining phases, while the *LSARA* shows the greatest values of *Un*(95) at 12.0267 and 12.995. The *LSGJA* achieved the smallest values by 10.717 and 10.03. The *OFU* scale, sometimes known as "The smaller, the most excellent," shows that *LSGJA* has the smallest value—3.3528, to be exact. This is nearly less than the values for *LSDMA*, which is 3.5985, and *LSARA*, which is 3.8142. Overall, the descriptions and justifications depict that all of the models perform greatly in the outline of estimation, with the superiority of *LSGJA* compared to *LSDMA*, followed by *LSARA*.

Moreover, Table 10.3 compares the better *LSGJA* algorithms' effectiveness measure (R^2) with the results of the $ICA-ANN$ method's literature [52]. Through comparison, it becomes clear that although the algorithms in the literature proved to be effective, the one developed in this research proved to be more accurate. A higher R^2

Table 10.3 Comparing the performance of *LSSVR* models

Stage	Index	*LSGJA*	*LSDMA*	*LSARA*	$ICA-ANN$ [52]
Train	R^2 (high get high point)	0.992	0.9915	0.99	0.9345
Test		0.9962	0.9936	0.9932	0.9556
Train	*RMSE* (low get high point)	3.8589	3.9969	4.3762	
Test		3.5996	4.5182	4.6612	
Train	*NRMSE* (low get high point)	0.0254	0.0263	0.0289	
Test		0.0251	0.0315	0.0326	
Train	*MAE* (low get high point)	2.9734	3.0081	3.128	
Test		2.6321	3.1406	3.1998	
Train	*TIC* (low get high point)	0.0122	0.0127	0.0139	
Test		0.0117	0.0147	0.0151	
Train	*IA* (high get high point)	0.998	0.9978	0.9974	
Test		0.999	0.9984	0.9983	
Train	*NMSE* (low get high point)	0.000715	0.00076	0.00084	
Test		0.00074	0.001	0.0011	
Train	*Un*(95) (low get high point)	10.7172	11.087	12.0267	
Test		10.0294	12.6055	12.995	
Train	*SI* (low get high point)	0.0254	0.0263	0.0288	
Test		0.0251	0.0315	0.0325	
OFU (low get high point)		3.3528	3.5985	3.8142	

value was attained throughout both the training and testing phases to accomplish this. For example, during the training phase, the R^2 values for *ICA* – *ANN* [52] are 0.9345, which is smaller than the R^2 value of 0.992 for *LSGJA*. In the same way, throughout the testing stages, *ICA* – *ANN* and *LSGJA* had R^2 values of 0.9556 and 0.9962, respectively.

Figure 10.5 displays the distribution of the simulated per recorded ratio (*SRR*) throughout both the training and testing phases, generated by the *LSGJA*, *LSDM*, and *LSARA* methods. A tighter range of *SRR*s with defined minimum and maximum limits indicates more accuracy. *LSARA* experiences reduced success in two distinct stages when there is a higher degree of variability in the *SRR*s. When employing the *GJA* approach to enhance the *LSSVR* approach, a significant occurrence of *SPR*s occurs when it is equal to one, and a narrower distribution is observed at *SRR* = 1, coupled with restricted upper and lower limits. The *LSDMA* analysis is located and operated between the mentioned algorithms, weaker than *LSGJA* and better than *LSARA*.

10.4.1 The Sensitivity Analysis

The main aim of sensitivity analysis (*SenAna*) is to understand how changes in input parameters or assumptions affect the output of a model or system. It is a technique used in various fields to assess the robustness and reliability of a model's results. To determine the impact of the input attributes on the target, a *SenAna* was conducted on the complete datasets. The *SenAna* for a given attribute (x_{in}) could potentially be calculated using the method below:

$$SenAna = \sum_{i=1}^{M} x_{in} y_i \Big/ \sqrt{\sum_{i=1}^{M} y_i^2 \sum_{i=1}^{M} x_{in}^2} \tag{10.48}$$

It's critical to keep in mind that the quantity *M* indicates the total number of data points, whereas the value y_i reflects the desired outcome. The numerical value of *M* in the context of this specific example is 156. Figure 10.6 is a graphical representation of the results that were obtained from the *SenAna*. Indicating the degree of connection that exists between each attribute and the objective that is expected, the value of *SenAna* may vary from 0 to 1, which reflects the degree of connection that occurs. This shows that the attribute parameter that has the most influence on the prediction's accuracy is the one that is being examined, as shown by a high value of *SenAna*, which is near 1. Figure 10.6 depicts that the simulation of *UCS* is highly affected by all attributes higher than 0.8435 (*CT*). The highest impact belonged to *MDD* at 0.9594, followed by *PH* at 0.9524.

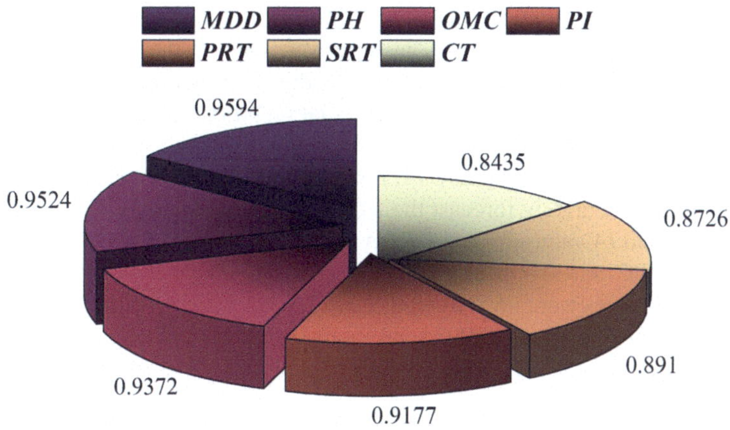

Fig. 10.6 The results of sensitivity analysis by *LSGJA*

10.4.2 Suggestions for Limitations

To improve the openness and credibility of the study, it is important to recognize and address certain constraints. The analysis is based on data obtained from previously released publications. The data's quality and trustworthiness may fluctuate across sources, and uncertainties may arise due to probable discrepancies in testing procedures or equipment employed by various researchers. The collection of data could contain constraints in terms of certain geographic regions, soil types, or experimental circumstances. The suggested model's capability to be applied to other areas or circumstances may be limited. Although the research utilizes state-of-the-art optimization techniques for hyperparameter tuning, it is important to recognize that the performance of these algorithms may be influenced by certain datasets or beginning circumstances, and their efficacy may differ. Unaccounted for factors like ambient conditions, loading patterns, or other variables not included in the research might impact the *UCS*. It is important to recognize these external elements as possible causes of uncertainty. It would be advantageous to provide prospective avenues for further investigation, such as examining the model's efficacy under varying soil conditions, integrating supplementary elements that influence the outcome, or performing field experiments to verify the model's predictions.

10.5 Conclusion

The present paper focused on estimating the *UCS* of *MC* modified with *RT* with novel, accurate, and reliable frameworks. For this aim, the powerful and dependable algorithms named least square support vector regression (*LSSVR*) analysis were considered by introducing percentage and size of recycled tiles (*PRT* and *SRT*), curing time (*CT*), maximum dry density (*MDD*), optimum moisture content (*OMC*), *PH*, and plasticity index (*PI*) of *MC* as input variables. In the present study, three cutting-edge optimization algorithms linked with *LSSVR* in order to tuning process, such as golden jackal algorithm (*GJA*), artificial rabbit optimizer (*ARA*), and dwarf mongoose algorithm (*DMA*).

The results indicate that the *LSGJA*, *LSARA*, and *LSDMA* methods have significant potential to accurately forecast the *UCS*. During the learning and examination stages, the R^2 and *IA* values for the *LSGJA* approach were found to be, respectively, 0.992 and 0.9962 and 0.998 and 0.999 (the greater, the better). With regard to *LSDMA*, these values were lower than those of *LSGJA*, with R^2 values of 0.9915 and 0.9936 and *IA* of 0.9978 and 0.9984, respectively. Despite having a strong prediction ability, the *LSSVR* algorithm optimized with *ARA* placed in third place by obtaining lower values.

Though it is more appropriate than *LSARA*, it is evident from the values of error-based measures that *LSGJA* performs more effectively in comparison with *LSDMA*. For instance, with regard to the values of the *TIC*, the slightest values owned by *LSGJA* at 0.0122 and 0.0117 in the learning and examining portions, respectively, slighter compared to *LSDMA* at 0.0127, and 0.0147, followed by 0.0139 and 0.0151 related to *LSARA*.

The *LSDMA* values for the uncertainty index are in the center for the learning and examining phases, while the *LSARA* shows the greatest values of *Un*(95) at 12.0267 and 12.995. The *LSGJA* achieved the smallest values by 10.717 and 10.03.

The *OFU* scale shows that *LSGJA* has the smallest value—3.3528, to be exact. This is nearly less than the values for *LSDMA*, which is 3.5985, and *LSARA*, which is 3.8142.

During the training phase, the R^2 values for $ICA - ANN$ [52] are 0.9345, which is smaller than the R^2 value of 0.992 for *LSGJA*. In the same way, throughout the testing stages, $ICA - ANN$ and *LSGJA* had R^2 values of 0.9556 and 0.9962, respectively.

The sensitivity analysis depicted that the simulation of *UCS* was highly affected by all attributes higher than 0.8435 (*CT*). The highest impact belonged to *MDD* at 0.9594, followed by *PH* at 0.9524.

Declarations

Conflict of Interests/Competing Interests The authors declare that they have no known conflict of interests/competing interests or personal relationships that could have appeared to influence the work reported in this paper.

Authors Contributions Both of the authors have contributed to all parts of the research.

Funding Not funded.

Data Availability The data that support the findings of this study are available on request from the corresponding author.

AI Use Disclosure Statement The authors did not use any generative AI or AI-assisted technologies in the creation of this manuscript. The authors used generative AI technologies only for proofreading. This includes minor corrections to grammar, spelling, and syntax. No AI was used for ideation, data analysis, or writing original content.

References

1. Al-Bared MAM, Marto A, Latifi N, Horpibulsuk S (2018) Sustainable improvement of marine clay using recycled blended tiles. Geotech Geol Eng 36:3135–3147
2. Al-Bared MAM, Marto A, Harahap ISH, Kasim F (2018) Compaction and plasticity comparative behaviour of soft clay treated with coarse and fine sizes of ceramic tiles. In: E3S web conference. EDP Sciences, p 1012
3. Vichan S, Rachan R (2013) Chemical stabilization of soft Bangkok clay using the blend of calcium carbide residue and biomass ash. Soils Found 53:272–281
4. Kim D, Park K (2013) An environmentally friendly soil improvement technology with microorganism. Int J Railw 6:90–94
5. Marto A, Latifi N, Eisazadeh A (2014) Effect of non-traditional additives on engineering and microstructural characteristics of laterite soil. Arab J Sci Eng 39:6949–6958
6. Saeed KA, Eisazadeh A, Kassim KA (2012) Lime stabilized Malaysian lateritic clay contaminated by heavy metals. Electron J Geotech Eng 17:1807–1816
7. Mahvash S, López-Querol S, Bahadori-Jahromi A (2017) Effect of class F fly ash on fine sand compaction through soil stabilization. Heliyon 3:e00274
8. Mirzababaei M, Arulrajah A, Ouston M (2017) Polymers for stabilization of soft clay soils. Procedia Eng 189:25–32
9. Saltan M, Fındık FS (2008) Stabilization of subbase layer materials with waste pumice in flexible pavement. Build Environ 43:415–421
10. Shahbazi M, Rowshanzamir M, Abtahi SM, Hejazi SM (2017) Optimization of carpet waste fibers and steel slag particles to reinforce expansive soil using response surface methodology. Appl Clay Sci 142:185–192
11. Ilieş N-M, Cîrcu A-P, Nagy A-C, Ciubotaru V-C, Kisfaludi-Bak Z (2017) Comparative study on soil stabilization with polyethylene waste materials and binders. Procedia Eng 181:444–451
12. Canakci H, Celik F, Bizne MOA, Bizne MOA (2016) Stabilization of clay with using waste beverage can. Procedia Eng 161:595–599
13. Elçi H (2016) Utilisation of crushed floor and wall tile wastes as aggregate in concrete production. J Clean Prod 112:742–752
14. Al Bakri AMM, Norazian MN, Kamarudin H, Ruzaidi CM (2008) The potential of recycled ceramic waste as coarse aggregates for concrete. In: Proc. MUCET, Malaysian univ. conf. eng. technol. Putra Brasmana, Perlis, pp 8–10
15. Al-Bared MAM, Marto A, Latifi N (2018) Utilization of recycled tiles and tyres in stabilization of soils and production of construction materials—a state-of-the-art review. KSCE J Civil Eng 22:3860–3874
16. Latifi N, Meehan CL, Abd Majid MZ, Horpibulsuk S (2016) Strengthening montmorillonitic and kaolinitic clays using a calcium-based non-traditional additive: a micro-level study. Appl Clay Sci 132:182–193
17. Al-Bared MAM, Marto A (2017) A review on the geotechnical and engineering characteristics of marine clay and the modern methods of improvements. Malays J Fundam Appl Sci 13:825–831
18. Shahri Z, Chan C-M (2015) On the characterization of dredged marine soils from Malaysian waters: physical properties. Environ Pollut 4:1
19. Oghli NK, Mahzad EF (2025) Predicting liquefaction triggering potential using metaheuristic GMDH approaches. Geotech Geol Eng 43:1–24. https://doi.org/10.1007/s10706-025-03275-z

20. Huat CY, Armaghani DJ, Bin Hashim H, Fattahi H, He X, Asteris PG, Fakharian P (2025) Development of a practical solution to predict surface settlement induced by twin tunnels. J Struct Des Constr Pract 30:4024097. https://doi.org/10.1061/JSDCCC.SCENG-1614
21. Mahzad E-F, Hooshang K, Meysam V, Jan A (2019) Predicting triaxial compressive strength and Young's modulus of frozen sand using artificial intelligence methods. J Cold Reg Eng 33:4019007. https://doi.org/10.1061/(ASCE)CR.1943-5495.0000188
22. Barkhordari MS, Armaghani DJ, Mohammed AS, Ulrikh DV (2022) Data-driven compressive strength prediction of fly ash concrete using ensemble learner algorithms. Buildings 12. https://doi.org/10.3390/buildings12020132
23. Esmaeili-Falak M, Sarkhani Benemaran R (2024) Ensemble extreme gradient boosting based models to predict the bearing capacity of micropile group. Appl Ocean Res 151:104149
24. Zhang K, Zhang Y, Razzaghzadeh B (2024) Application of the optimal fuzzy-based system on bearing capacity of concrete pile. Steel Compos Struct 51:25–41. https://doi.org/10.12989/scs.2024.51.1.025
25. Motamedi S, Shamshirband S, Petković D, Hashim R (2015) Application of adaptive neuro-fuzzy technique to predict the unconfined compressive strength of PFA-sand-cement mixture. Powder Technol 278:278–285
26. Esmaeili-Falak M, Letafat A (2025) Prediction of unconfined compressive strength in pozzolanic geopolymer-stabilized granular materials using tree-based models. Transp Infrastruct Geotechnol 12:247. https://doi.org/10.1007/s40515-025-00701-w
27. Gullu H (2017) On the prediction of unconfined compressive strength of silty soil stabilized with bottom ash, jute and steel fibers via artificial intelligence. Geomech Eng 12:441–464
28. Li D, Zhang X, Kang Q, Tavakkol E (2023) Estimation of unconfined compressive strength of marine clay modified with recycled tiles using hybridized extreme gradient boosting method. Constr Build Mater 393:131992. https://doi.org/10.1016/j.conbuildmat.2023.131992
29. Shamshirband S, Tavakkoli A, Roy CB, Motamedi S, Song K-I, Hashim R, Islam SM (2015) Hybrid intelligent model for approximating unconfined compressive strength of cement-based bricks with odd-valued array of peat content (0–29%). Powder Technol 284:560–570
30. Tabarsa A, Latifi N, Osouli A, Bagheri Y (2021) Unconfined compressive strength prediction of soils stabilized using artificial neural networks and support vector machines. Front Struct Civil Eng 15:520–536
31. Mozumder RA, Laskar AI, Hussain M (2017) Empirical approach for strength prediction of geopolymer stabilized clayey soil using support vector machines. Constr Build Mater 132:412–424
32. Suthar M (2020) Applying several machine learning approaches for prediction of unconfined compressive strength of stabilized pond ashes. Neural Comput Appl 32:9019–9028
33. Motamedi S, Shamshirband S, Hashim R, Petković D, Roy C (2015) Retracted: Estimating unconfined compressive strength of cockle shell–cement–sand mixtures using soft computing methodologies. Eng Struct 98:49
34. Van Giap L, Pham TA, Vu H-LT (2022) Application of random forest model to determine unconfined compressive strength of the soil-cement mixture. In: CIGOS 2021, emerging technologies and applications for green infrastructure, pp 1775–1783
35. Wang M, Wan W, Zhao Y (2020) Prediction of the uniaxial compressive strength of rocks from simple index tests using a random forest predictive model. C R Méc 348:3–32
36. Pham BT, Qi C, Ho LS, Nguyen-Thoi T, Al-Ansari N, Nguyen MD, Nguyen HD, Ly H-B, Van Le H, Prakash I (2020) A novel hybrid soft computing model using random forest and particle swarm optimization for estimation of undrained shear strength of soil. Sustainability 12:2218
37. British Standards Institution (1990) BSI methods of test for soils for civil engineering purposes: part 2, classification tests, London
38. British Standards Institution (1990) BSI 1377 methods of test for soils for civil engineering purposes: part 4, compaction related tests, London, Milton Keynes
39. Al-Bared MAM, Harahap ISH, Marto A (2018) Sustainable strength improvement of soft clay stabilized with two sizes of recycled additive. GEOMATE J 15:39–46

40. Al-Bared MAM, Marto A (2017) Evaluating the compaction behaviour of soft marine clay stabilized with two sizes of recycled crushed tiles. In: Global civil engineering conference. Springer, pp 1273–1284
41. Narendra BS, Sivapullaiah PV, Suresh S, Omkar SN (2006) Prediction of unconfined compressive strength of soft grounds using computational intelligence techniques: a comparative study. Comput Geotech 33:196–208. https://doi.org/10.1016/j.compgeo.2006.03.006
42. Cohen I, Huang Y, Chen J, Benesty J, Benesty J, Chen J, Huang Y, Cohen I (2009) Pearson correlation coefficient. In: Noise reduction in speech processing. Springer, pp 1–4
43. Chopra N, Mohsin Ansari M (2022) Golden jackal optimization: a novel nature-inspired optimizer for engineering applications. Expert Syst Appl 198:116924. https://doi.org/10.1016/j.eswa.2022.116924
44. Agushaka JO, Ezugwu AE, Abualigah L (2022) Dwarf Mongoose optimization algorithm. Comput Methods Appl Mech Eng 391:114570. https://doi.org/10.1016/j.cma.2022.114570
45. Wang L, Cao Q, Zhang Z, Mirjalili S, Zhao W (2022) Artificial rabbits optimization: a new bio-inspired meta-heuristic algorithm for solving engineering optimization problems. Eng Appl Artif Intell 114:105082. https://doi.org/10.1016/j.engappai.2022.105082
46. Vapnik V, Vapnik V (1998) Statistical learning theory, vol 1. Wiley, New York, p 2
47. Armaghani DJ, Yang P, He X, Pradhan B, Zhou J, & Sheng D (2024) Toward Precise Long-Term Rockburst Forecasting: A Fusion of SVM and Cutting-Edge Meta-heuristic Algorithms: DJ Armaghani et al. Natural Resources Research, 33(5), 2037–2062
48. Asteris PG, Karoglou M, Skentou AD, Vasconcelos G, He M, Bakolas A, ... & Armaghani DJ (2024) Predicting uniaxial compressive strength of rocks using ANN models: incorporating porosity, compressional wave velocity, and schmidt hammer data. Ultrasonics, 141, 107347
49. He B, Armaghani DJ, Lai SH, He X, Asteris PG, & Sheng D (2024) A deep dive into tunnel blasting studies between 2000 and 2023—A systematic review. Tunnelling and Underground Space Technology, 147, 105727
50. Khandelwal M, Armaghani DJ, Faradonbeh RS, Yellishetty M, Majid MZA, & Monjezi M (2017) Classification and regression tree technique in estimating peak particle velocity caused by blasting. Engineering with Computers, 33(1), 45–53
51. Suykens JAK, De Brabanter J, Lukas L, Vandewalle J (2002) Weighted least squares support vector machines: robustness and sparse approximation. Neurocomputing 48:85–105. https://doi.org/10.1016/S0925-2312(01)00644-0
52. Al-Bared MAM, Mustaffa Z, Armaghani DJ, Marto A, Yunus NZM, Hasanipanah M (2021) Application of hybrid intelligent systems in predicting the unconfined compressive strength of clay material mixed with recycled additive. Transp Geotech 30:100627

[redacted]

The manufacturer's authorised representative in the EU is Springer Nature Customer Service Centre GmbH, Europaplatz 3, 69115 Heidelberg, Germany. If you have any concerns regarding our products, please contact ProductSafety@springernature.com

Printed and bound by CPI Group (UK) Ltd, Croydon, CR0 4YY
21/09/2026
02220015-0004